T0190118

Sustainable Textiles: Production, Processing, Manufacturing & Chemistry

Series Editor

Subramanian Senthilkannan Muthu, Head of Sustainability, SgT and API, Kowloon, Hong Kong

This series aims to address all issues related to sustainability through the lifecycles of textiles from manufacturing to consumer behavior through sustainable disposal. Potential topics include but are not limited to: Environmental Footprints of Textile manufacturing; Environmental Life Cycle Assessment of Textile production; Environmental impact models of Textiles and Clothing Supply Chain; Clothing Supply Chain Sustainability; Carbon, energy and water footprints of textile products and in the clothing manufacturing chain; Functional life and reusability of textile products; Biodegradable textile products and the assessment of biodegradability; Waste management in textile industry; Pollution abatement in textile sector; Recycled textile materials and the evaluation of recycling; Consumer behavior in Sustainable Textiles; Eco-design in Clothing & Apparels; Sustainable polymers & fibers in Textiles; Sustainable waste water treatments in Textile manufacturing; Sustainable Textile Chemicals in Textile manufacturing. Innovative fibres, processes, methods and technologies for Sustainable textiles; Development of sustainable, eco-friendly textile products and processes; Environmental standards for textile industry; Modelling of environmental impacts of textile products; Green Chemistry, clean technology and their applications to textiles and clothing sector; Eco-production of Apparels, Energy and Water Efficient textiles. Sustainable Smart textiles & polymers, Sustainable Nano fibers and Textiles; Sustainable Innovations in Textile Chemistry & Manufacturing; Circular Economy, Advances in Sustainable Textiles Manufacturing; Sustainable Luxury & Craftsmanship; Zero Waste Textiles.

More information about this series at http://www.springer.com/series/16490

Miguel Ángel Gardetti ·
Subramanian Senthilkannan Muthu
Editors

Handloom Sustainability and Culture

Product Development, Design and Environmental Aspects

Editors
Miguel Ángel Gardetti
Center for Studies on Sustainable Luxury
Buenos Aires, Argentina

Subramanian Senthilkannan Muthu
SgT Group and API
Hong Kong, Kowloon, Hong Kong

ISSN 2662-7108 ISSN 2662-7116 (electronic)
Sustainable Textiles: Production, Processing, Manufacturing & Chemistry
ISBN 978-981-16-5667-5 ISBN 978-981-16-5665-1 (eBook)
https://doi.org/10.1007/978-981-16-5665-1

This Springer imprint is published by the registered company Springer Nature Singapore Pte Ltd.
The registered company address is: 152 Beach Road, #21-01/04 Gateway East, Singapore 189721, Singapore

Preface

The title *Sustainability, Culture and Handloom* consisting of three volumes—Entrepreneurship, Culture and Luxury; Artisanship and Value Addition; and Product Development, Design and Environmental Aspects—arises from an idea originated on 7 August 2020 on the occasion of the National Handloom Day in India. It is held annually to commemorate the Swadeshi Movement launched in 1905. It is a tribute that publishers do in order to celebrate this day.

Artisanship and craftmanship bring a whole new universe of small details and differences in garments that stand out between the common homogenized products characteristics of consumerist society. These pieces, which hold the essence of heritage and uniqueness through techniques, materials and handmade products, might be defined as "cultural luxury". Beyond the essence and scarcity in handicraft products, these pieces are associated with an emotional value precisely due to the cultural process of maintaining cultural heritage based on traditions that are passed from generation to generation [1]. Craftsmanship luxury pieces carry the evidence of knowledge and ancient techniques that passes through generations, assuring cultural diversity and heritage, and those are intangibles of artisanship.

Publishers and authors of this title are moved by the motivation to preserve and pass on the savoir-fair of those wise hands that ensure ancient techniques. They intend to preserve this knowledge and pass it over through generations.

Social and cultural aspects in craftmanship reflect and keep alive the essence and hold the values, the traditions and the cultural social exchange. The cultural sustainability of handicrafts is also, to keep this knowledge applied, and present. For example, demonstrating through the use of handicrafts in our daily lives, and keeping the spirit of artisanship in mind. In summary, envisioning and expanding the traditions and culture expressed in the artisan pieces assure, maintain and sustain cultural heritage and diversity [2].

The volume *Product Development, Design and Environmental Aspects* consists of five chapters that give a comprehensive outlook on this subject and begins with the work titled "Environmental sustainability of handloom sector" by Bhagyashri N. Annaldewar, Nilesh C. Jadhav and Akshay C. Jadhav. In this chapter, they highlight the sustainability criteria of the handloom sector which tends to focus on the environmental aspects regarding textile processes. They analyse waste water during production, greenhouse gas emissions and chemical pollution.

The following chapter, "Sustainability, Culture and Handloom Product Diversities with Indian Perspective" written by Dr. Sanjoy Debnath, explores environmental aspects at every step involved in raw material preparation for handloom, handloom weaving and post-production of fabrics and products. The author presents an overview of sustainability of Indian handloom industry, its products and cultural interventions.

Then, Dorina Horătău developed the chapter titled "Teaching about 'fibre', between art and contemporary design". This research proposes a teaching method to focus on the symbolism of craftsmanship in handlooms. It analyses the art and the potential symbols of craftsmanship, and how it can be transferred.

Subsequently, the purpose of Dr. Jenny Pinski, Dr. Faith Kane and Dr. Mark Evans in "Handweaving as a catalyst for sustainability" is to explore the role of craft, craft-based design and handweaving in relation to key models of sustainability and the circular economy. Through a case study, it exemplifies and analyses the potentials of circular economy in handweaving handcrafts.

Finally, the last chapter, "Handloom—The Challenges and Opportunities"—by the author Sankar Roy Maulik—presents handlooms in India, and analyses challenges and potentials as a job generator in rural areas and vulnerable sectors. It focuses on environmental aspects, sustainability and value creation of artisanal handcraft pieces that can improve the livelihood of the artisans.

Buenos Aires, Argentina — Miguel Ángel Gardetti
Hong Kong — Subramanian Senthilkannan Muthu

References

1. Guldager S (2015) Irreplaceable luxury garments. In: Gardetti MA, Muthu SS (eds) Handbook of sustainable luxury textiles and fashion, vol 2. Springer, Singapore, pp 73–97
2. Na Y, Lamblin M (2012) Sustainable Luxury: Sustainable Crafts in a Redefined Concept of Luxury from Contextual Approach to Case Study Making Futures Journal Vol 3 ISSN 2042-1664

Contents

Environmental Sustainability of Handloom Sector 1
Bhagyashri N. Annaldewar, Nilesh C. Jadhav, and Akshay C. Jadhav

Sustainability, Culture and Handloom Product Diversities with Indian Perspective 23
Sanjoy Debnath

Teaching About "Fibre": Between Art and Contemporary Design 49
Dorina Horătău

Handweaving as a Catalyst for Sustainability 77
Jenny Pinski, Faith Kane, and Mark Evans

Handloom—The Challenges and Opportunities 97
Sankar Roy Maulik

About the Editors

Miguel Angel Gardetti Ph.D. founded the **Centre for Study of Sustainable Luxury**, first initiative of its kind in the world with an academic/research profile. He is also the founder and director of the "Award for Sustainable Luxury in Latin America". For his contributions in this field, he was granted the "**Sustainable Leadership Award** (academic category)", in February 2015 in Mumbai (India). He is an active member of the **Global Compact** in Argentina—which is a **United Nations** initiative—and was a member of its governance body—Board of The Global Compact, Argentine Chapter—for two terms. He was also part of the task force that developed the **Management Responsible Education Principles** of the United Nations Global Compact. This task force was made up of over 55 renowned academics worldwide pertaining to top Business Schools.

Dr. Subramanian Senthilkannan Muthu currently works for SgT Group as Head of Sustainability, and is based out of Hong Kong. He earned his PhD from The Hong Kong Polytechnic University, and is a renowned expert in the areas of Environmental Sustainability in Textiles & Clothing Supply Chain, Product Life Cycle Assessment (LCA) and Product Carbon Footprint Assessment (PCF) in various industrial sectors. He has 5 years of industrial experience in textile manufacturing, research and development and textile testing, and over a decade of experience in life cycle assessment (LCA), carbon and ecological footprints assessment of various consumer products. He has published more than 100 research publications, written numerous book chapters and authored/edited over 100 books in the areas of Carbon Footprint, Recycling, Environmental Assessment and Environmental Sustainability.

Environmental Sustainability of Handloom Sector

Bhagyashri N. Annaldewar, Nilesh C. Jadhav, and Akshay C. Jadhav

Abstract Environmental sustainability is one of the most important assets of any system which should be diverse and yet should be productive for the growth of any industry. The environmental sustainability can serve many solutions to the handloom sector financially by recycling the textile waste using proper disposal methods and attenuation of textile products used in the handloom sector without harming the environment. The main aim is to generate environmentally sustainable method to these numerous problems that take place in handloom sector. To quantify potential ecological benefits and economic effects of handloom sector can be optimized by implementing textile waste recycling, minimizing the usage of virgin materials, less power consumption or usage of green energy, less water consumption, usage of eco-friendly dyes and materials, etc. The current chapter highlights the sustainability criteria of handloom sector which tends to focus on the environmental aspect regarding textile processing. Further, it also deals with more thorough information regarding the waste water production, greenhouse gas emissions and chemical pollution which is associated with the handloom sector. It would also allow more comprehensive assessment on the environmental sustainability in handloom sector.

Keywords Handloom · Environment · Sustainability · Textiles · Recycle

1 Introduction

Handloom industry is probably the utmost prehistoric industry throughout the entire existence of human civilization. Like air, food, water and housing, accessibility of adequate clothing is prime need of life. The clothing stands first in list of important materials of man when he shows up on the incredible stage of the world. A recently conceived infant is promptly wrapped up in a clean and delicate fabric, to shield him from rough climate. A piece of fabric secures him as a shell ensures a glittering or

B. N. Annaldewar · N. C. Jadhav · A. C. Jadhav (✉)
Department of Fibres and Textile Processing Technology, Institute of Chemical Technology, University under section-3 of UGC act 1956, Mumbai 400019, India

M. Á. Gardetti et al. (eds.), *Handloom Sustainability and Culture*, Sustainable Textiles: Production, Processing, Manufacturing & Chemistry,
https://doi.org/10.1007/978-981-16-5665-1_1

sparkling pearl. In fact, his apparel is his closest, best and 'mobile' cover. Along these lines, the textile business has central significance in our life.

In primal age, man couldn't have abandoned garments and civilization itself starts with the garments. Weaving is the most significant of all creative art. 1/5th of working world is occupied with weaving and its different branches. Present-day culture is no less obligated to the respectable and stately workmanship, which has been purported from a king to and ordinary man in various times of history.

Hand weaving has been the essential action of human culture since ancient days in which utility and aesthetics are mixed together. It is said that, 'one who works with one's hands is a labour; one who works with one's hands and brain is a craftsman and one who works with one's hands and heart is an artist'. Handloom weavers are for sure a band of imaginative craftsmen as they pour their whole souls in to the work. The weavers weave with yarn as well as their profound sentiments and feelings are additionally woven in the surface. The 'hand-woven' fabric is representative of man's endeavour to bring magnificence and elegance into a day-to-day existence, which is generally seriously obliged by the standardization and the consequent monotony [1].

In any case, in the current context of globalization and fast innovative changes, handloom area is assailed with numerous difficulties and the handloom items are being reproduced on power looms at much lower cost. The generous upgrades in human prosperity resulting because of more prominent control and exploitation of normal resources, generally food, fibre, fuel, timber and water have cost more broad biological system harm in the previous 50 years than any time in our history. By suggestion, old mechanical models of overseeing of the world's biological systems seem to have fizzled, requiring re-established endeavours to at the same time relieve the current causes while creating defensive answers for what's to come. Set-up conceptual solutions for the sustainability issue seem to bode well—a blend of proficiency, limitation and repair. People need to direct destructive practices to accomplish a worthy harmony between the speed with which we transform natural resources and the rate the biosphere can restore them, while satisfying needs of an expanding population, a significant number of whom need and aspire improved standards of living [2]. In the recent years, developing cognizance about ecological conservations and control of pollution has recharged interest in eco-friendly items. Eco-friendly textiles and garments are the popular buzz nowadays. Designers, manufacturers and retailers are occupied and busy to create 'green' product range for the mass market [3].

This chapter gives an outline about environmental sustainability in handloom industry. It assists with understanding the idea of sustainability and gives different solutions for environmentally sustainable improvement in handloom [4, 5].

2 Handloom Fabric Manufacturing Process

Figure 1 shows the processes involved in handloom fabric manufacturing which are briefly explained below:

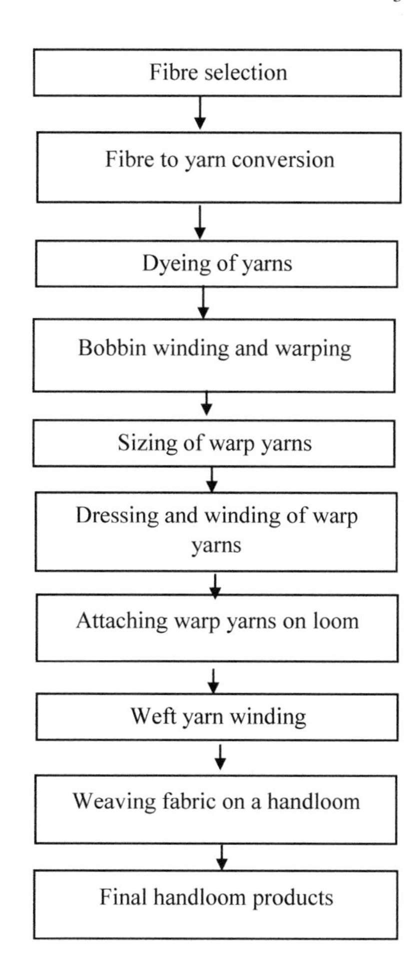

Fig. 1 Processes involved in handloom fabric manufacturing

- **Raw materials**

Natural fibres like cotton, wool, linen, jute and silk are the most common picks of raw material which are used in handloom weaving.

- **Raw material to yarn conversion**

The raw material is delicately moved with palm to create an approximately cylindrical bunch which is known as a sliver. Further, this loosely interlocked sliver is spun on a spinning wheel to shape a dense and fine yarn. Later, the spun yarns are interlaced into skeins and sent for dyeing. Nowadays, mill spun yarns are used by handloom weavers.

- **Dyeing of yarns**

Prior to the dyeing, yarns are scoured to remove natural impurities. Dyeing of the yarn is carried out by hand in the form of hanks using natural or synthetic colourants.

- **Bobbin winding and warping**

With the assistance of turning wheel, coloured yarn hank gets changed over into a linear thread structure and wounded on the bobbin. This procedure allows arranging yarn lengths for weaving. Further, warping is done, which is the parallel arrangement and winding of warp yarn from bobbin to the warp beam. Customarily, the weavers utilize a huge rotating drum as warp beams and decide the width and length of the fabric. These drums help them in checking the number and colour-wise grouping of yarns. Additionally, the dimensions of the warp are chosen by the weaver at this stage.

- **Sizing of warp yarns**

Before the post-warping is done, the warp yarns (made of cotton and linen) are overextended for the application of size. Sizing material is applied to add strength and lubricate the yarn.

- **Dressing and winding the warp yarns**

Before applying the sizing, material warp yarns are stacked onto the loom, the warp yarns are adjusted and isolated to encourage smooth weaving. The sized yarns are aligned properly and is cautiously wounded around a wooden beam and conveyed to the loom.

- **Attaching warp yarns on loom**

Every warp yarn is drawn through heddles and reed lastly tied on both front beam and back beam. As indicated by a pre-decided weave plan, yarns are passed through heddles that separate the warp yarns into two sections between which the weft yarn passes.

- **Weft yarns winding**

A hank of yarn is wound onto a little bobbin called ‘pirn’. The weft yarn wounded on pirn is then embedded into a shuttle.

- **Weaving fabric in a handloom**

As the name proposes, handloom is a loom that is utilized to weave fabrics utilizing hands, that is, without the utilization of power. Handloom fabric is woven by entwining the warp (length-wise thread) and weft (width-wise thread). The warp threads move vertically in up–down motion to form shed. The shuttle passes through the horizontal thread and the movable comb-type frame beats the woven fabric. The heddle then shifts which sets the warp in the opposite direction and this binds the weft. The loom is the basic equipment of hand weaving. Two kinds of looms are used, viz. pit looms and frame looms. Pit looms are used in making coloured fabrics such as towels, bedsheets, handkerchiefs, etc. Frame looms on the other hand are used to make designed fabrics such as heavy woven bedsheets, striped and checked material, gauze cloth, etc. Weavers usually continue weaving for extended hours in a day which involves massive concentration and physical strength [6, 7].

3 Impact of Handloom Industry on Environment

Natural fibres used for handloom fabric preparation have several impacts on environment. For example, cotton fibre which is widely used for handloom fabric producing don't merit the natural fibre name as more than 99% of all cotton is produced utilizing chemical and synthetic fertilizers to get better return. Cotton crop is harmed majorly by 230 types of insects everywhere on the world. Because of its weakness to pests and insects, cotton is the significant consumer of agrochemicals on the planet. 16% of the world's insecticides are utilized for the single cotton crop. The utilization of different poisonous pesticides and synthetic fertilizers bring about upsetting the natural habitat, expanding the expense of production, advancement of insect species which are impervious to insecticide sprays and changing the insect pattern. Other fundamental ecological consequences are the water bodies and air contamination, lessening of biodiversity and disturbing the environment [8]. Then again, silk is an inexhaustible material with low environmental impact compared with respect to other fibres. The silkworms feed on mulberry leaves, which don't warrant a lot of utilization of pesticides or fertilizers to grow. The primary issue in silk production is the executing of the larvae when the cocoon is boiled to separate the filament during the sericulture process; this is disapproved intensely by numerous animal welfare government assistance and rights activists [9].

Processing of fibres previously and during the spinning and weaving operations generates dust and lint, which harms the workplace of industry. Dust may prompt respiratory diseases among the labourers [10]. Air pollution is additionally caused during dyeing and printing of textiles because of steam generated by coal and water. At the point when the steam is generated, it produces carbon, carbon dioxide, carbon monoxide and sulphur, which again cause air pollution. It perilously affects the health of individuals and animals which may result in mortality to eye, respiratory issues, reducing perceivability, persistence of fog and so on [11].

Since the handloom processing industry creates monstrous quantity of solid and liquid waste, it has been pondered as probably the biggest polluter of the environment. Handloom processing sector is a huge source of poisonous effluents by its real nature, as it requires gigantic amount of water for its different processing steps. Textile effluents are synthetically complex in nature, having a higher load of colours, organic salts, pH, solvents and heavy metals. The discharge of effluents into the environment changes the smell, colour and composition of surface water and groundwater and along these lines seriously influences widely varied flora and fauna [12].

Handloom processing sector is a huge source of poisonous effluents by its real nature, as it requires gigantic amount of water for its different processing steps. Textile effluents are synthetically complex in nature, having a higher load of colours, organic salts, pH, solvents and heavy metals. The discharge of effluents into the environment changes the smell, colour and composition of surface water and groundwater.

Numerous pre-loom and post-loom processes are engaged with the production of handloom fabric, for example, dyeing of yarn, sizing, winding, warping, preparation of beam, etc., are some of the pre-loom actions, whereas bleaching, dyeing, printing, finishing and calendaring are the post-loom activities, which must be done in a successive way. The weaving on handloom is a dreary just as repetitive nature of work which requires continuous endeavours.

The fundamental explanations behind noise pollution might be because of high friction between the moving parts and the kind of material of the moving parts in the handloom. Also, when there is work pressure, the noise level is prevailing. Despite the fact that the weavers weave in the rhythmic development, they as often as possible ignore the power of sound level delivered yet it is accidentally influencing their auditory framework [13].

Disposal of waste is quite possibly the most genuine environmental issues the public is facing. Incineration of waste and dumping of waste in landfills have negative impact on environment. Natural fibres which are utilized for handloom fabric, for example, cotton, jute, wool, linen and silk are biodegradable. Nonetheless, decomposition of biodegradable materials in landfills is a major reason for methane, a significant greenhouse harming gaseous substance, making landfills one of the biggest human-related sources of methane. Likewise, incinerating waste can cause CO_2 emissions and remaining ashes may contain harmful substances [14].

Carbon footprints and its minimization is a significant subject nowadays and all of us needs to lessen the carbon footprints to the most extreme conceivable degree to protect our living planet. A carbon footprint is a proportion of measure of greenhouse gaseous substances created through burning of petroleum products for electricity, heating and transportation and so forth. Clothing, being one of the essential requirements of people, makes carbon footprint in each phase of the existence life cycle of a textile material. Textiles and clothing industry include a lengthy and complicated store supply chain, which is accountable for the huge measure of carbon footprint generation and it is one of the central sources of emanations of greenhouse gaseous substances. Production, transportation, utilization and disposal of textile materials represent a danger to carbon footprint [15].

4 Sustainability

Sustainability implies an ability to keep up some entity, result or process after some time. Sustainability can likewise be characterized as the productive and impartial dissemination of assets intra-generationally and between generationally with the activity of financial exercises inside the limits of a limited environment [16]. Sustainability includes the mix of ecological well-being, social value and monetary essentialness to make flourishing, sound, various and strong networks for this age and ages to come. The act of sustainability perceives how these issues are interconnected and requires a frameworks approach and an affirmation of intricacy.

The most acknowledged definition of sustainable advancement is the improvement that addresses the issues of the current age without compromising off the capacity of people in the future to addresses their own issues [17].

A sustainable society needs to meet three conditions: its rates of utilization of inexhaustible assets ought not to surpass their paces of recovery; its paces of utilization of non-sustainable assets ought not to surpass the rate at which reasonable inexhaustible substitutes are created; and its paces of contamination of outflow ought not to surpass the assimilative limit of the environment [18].

4.1 Aspects of Sustainability

There are three aspects of sustainability which are discussed below:

- **Social**

The social component of sustainability depends on the way that equivalence and understanding of the interdependence of individuals within the society are the basic requirement for an adequate quality of life, which is the principal objective of development [19].

The social aspect of sustainability refers to, in wide terms, public strategies that help social issues. These social problems relate to our prosperity and incorporate features like health care, housing, education, employment, etc., and so forth it guarantees that people do approach social administrations, don't endure lack of information on their privileges and exercise a dependable impact on the advancement of social strategies and amenities, both locally and nationally [20].

- **Environmental**

Environment can be characterized as the physical encompassing of man/woman of which he/she is a part on which he/she is dependent for his/her exercises like physiological working, creation and utilization. His actual climate environment from air, water and land to natural resources. Environmental degradation is an intense issue overall which covers an assortment of issues including contamination, biodiversity misfortune and animal extinction, deforestation and desertification and global warming and much more. The environmental degradation and deterioration of the environment through diminution of resources incorporates all the biotic and abiotic elements that structure our surrounding that is water, air, soil, plant, animals and any other living and non-living component of the earth. The main consideration of environment degradation is human (present day urbanization, industrialization, overpopulation development, deforestation and so on) and natural calamities like (hurricanes, floods, rising temperatures, dry seasons, fires, etc.)

Environmental pollution alludes to the degradation of the quality and amount of natural assets. Various types of human exercises are the fundamental reasons behind environmental degradation. For instance, the smoke transmitted by the vehicles and processing industries extends the proportion of poisonous gases which is recognizable all around. The waste things, smoke transmitted by vehicles are the central driver of pollution. Unconstrained urbanization and industrialization have caused air, water and sound pollution. Urbanization and industrialization help to grow pollution of the sources of water. So additionally, the smoke released by vehicles and substances like carbon monoxide, chlorofluorocarbon, nitrogen oxide and other clean elements cause air contamination [21, 22].

The harm caused by the human beings to the environment is right now not included as an expense in monetary and social terms. This absence of 'environmental value' has permitted us to over-misuse 'free' natural resources—which are, obviously, not free. It has additionally prompted over-production of cheap products with exceptionally short life expectancies which are generously disposed of into the environment after utilizing and afterwards new cheap merchandise are bought and disposed of again and this cycle continues endlessly influencing the planet's ability to re-establish its ecological administrations eventually [23].

The idea of environmental sustainability is about the natural habitat and how it stays profitable and strong to help human existence. Environmental sustainability relates to biological system honesty and conveying limit of natural habitat. Environmental sustainability improves human well-being assistance by securing resources of crude materials utilized for human necessities and guarantees that sinks for human squanders are not surpassed to forestall damage to people. The ramifications are that natural resources should be collected no quicker than they can be recovered, while waste should be produced no quicker than they can be acclimatized by the environment. This is on the grounds that the earth frameworks have limits inside which balance is maintained [16, 24].

More than 90% of the emissions are created from the five activities in the textile and clothing production that are bleaching, dyeing, finishing, weaving and fibre production (Refer Fig. 2). Likewise, the textile and clothing sector has created in excess of 2 billion tons of carbon dioxide which addresses 4% of worldwide global carbon emissions. United Nations Climate Change News states that the Textile and fashion industry contribute 10% of the overall greenhouse gas emissions because of its long supply chains and energy escalated production. Additionally, almost 20–22% of waste water is used by the textile and clothing sector. Textile industry is one of

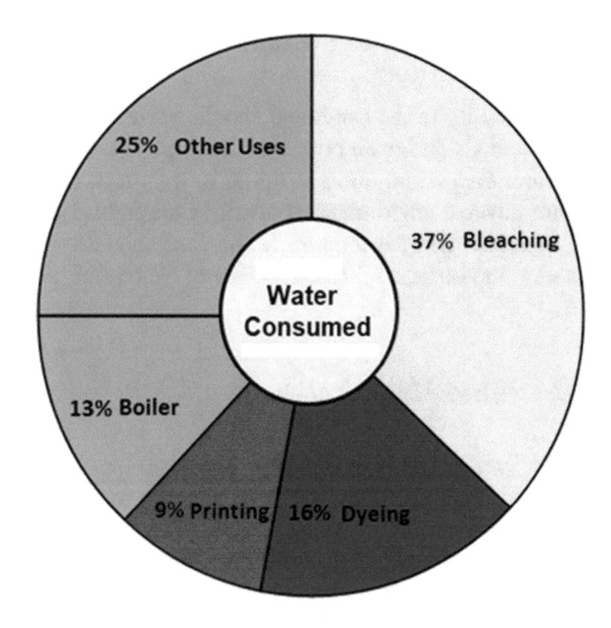

Fig. 2 Water consumed during wet processing in handloom units

the top three water wasting industry on the planet by utilizing over 2.5 billion tons of water every year. Cotton cultivating uses practically 24% of the insecticides and 11% of pesticides despite using just 3% of complete overall cultivable land and right around 20,000 L of water is required to produce 1 kg of cotton [25].

- **Economic**

The centre necessity of sustainability is that current financial exercises would not bring about an exorbitant burden on people in the future. This rule is sufficiently enough to suggest distinctive choice principles for the protection. Sustainability advancement requires upkeep of the natural and human resource base that is fundamental for long haul financial development of ecological resources [26].

Economic sustainability is utilized to characterize different techniques that promote the use of financial resources for their best potential benefit. A sustainable economic model recommends a fair dispersion and proficient designation of resources. The thought is to promote the utilization of those assets in an effective and capable manner that gives long-term benefits and sets up productivity. The pleasant thing about adopting an all-out strategy to sustainability is that if you focus around social and ecological issues, benefits will frequently follow. Social activities affect the consumer behaviour and worker performance, while environmental activities, for example, energy productivity and contamination mitigation can directly affect diminishing waste.

Economic sustainability ensures the business makes a benefit, yet additionally that business tasks don't make social or environmental issues that would hurt the long-term achievement of the organization [27].

5 Environmental Sustainability of Handloom Sector

Sustainability in the handloom should intent to maintain ecologically safe production methods during all phases of development while balancing environmental equilibrium. Paramount aim should be to minimize the usage of toxic chemicals that cause adverse environmental effect in the industry. Alongside it also should seek to decrease the consumption of the water and energy [28]. Various approaches to achieve environmental sustainability in handloom sector are mentioned below.

5.1 Waste Minimization

Waste from handloom sector is delegated pre-customer and post-shopper waste. Pre-consumer waste comprises of by-product materials generated during the preparation of yarn and fabric, for example, fibres, yarns, fabric waste and so forth Post-buyer waste primarily originates from household unit sources and comprises of pieces of clothing or textiles that the proprietor at this point don't needs and disposes them of [29].

At the point when a textile material is discarded as post-consumer waste in a landfill, the entire materials and energy which are utilized during its production as well as the carbon emission by-products from transport of the item along the production network and the supply chain implemented by labour contribution all through these stages are wasted. Most nations are attempting to diminish the measure of removal in landfills. Wastes management comprises of the different stages, for example, reuse, recycling and, last but not least, landfilling [30]. Waste minimization approaches are examined and discussed below:

- **Reuse**

Product reuse is a non-damaging cycle that finds a second or further use for end-of-first life materials without its change in state, in this way rescuing or restoring a conceivably disposed of thing into dynamic use. Reuse plays a very critical part in the minimization of waste as imagined by waste counteraction and zero waste procedures [31]. Presently, not many utilized textile or clothing shops have been set up which will offer alternative option in contrast to utilized garments, without being discarded; it tends to be re-utilized and sold at less expensive price. Individuals generally give their old garments to government-assisted welfare associations which not just assists to minimize the burden of the destitute peoples but also in addition it helps in space management at dumping places and environmental pollution problems.

- **Recycling**

The meaning of recycle means that discarded materials are repaired, re-designed or re-utilized in different conditions. Recycling means improving the process of material utilization in any manners, for example, waste being transformed into raw materials for nature or some other industry.

Recycling activities are among the exercises that contribute in the direction of sustainability of the environment that can diminish decimation of nature and decrease the use of crude materials to make new things which is effective in the light of the fact that it tends to be utilized on various occasions before disposal. Recycling process can decrease waste, diminish utilization of new crude materials, lessen use of energy, decrease pollution (water and air) and decrease creation of ozone harmful gases [32]. For textiles and garments materials which are inadmissible for reuse, recycling into new fibres, feedstocks, materials and items should be possible. Recycling frequently comprises of a blend of mechanical, chemical and thermal process. Scarcely some examples are given below:

- Conversion of post-buyer clothing materials into wipers. Wiper creation includes eliminating any metallic parts, for example, zips from waste materials of clothing prior to cutting into the significant size available for sale to industry.
- Post-shopper and pre-purchaser materials can likewise be utilized as feedstock in the creation of fibres for filling materials, for example, flocking, insulation purpose, nonwovens and in the creation of shoddy fibre filaments for recycled yarns.
- For these reasons, textile materials are mechanically recycled in processes which cut and shred fabrics into filaments which would then be able to be respun into yarn or made into nonwoven materials. Flocking is comprised of shredded textile materials combined with wool to make a flame-retardant product and is utilized for filling in mattresses and furnishing goods. Nonwovens are made from bonded layers of shredded fabric material and utilized for purposes, such as insulation protection and carpet underlay [33].

- **Upcycling**

Recycling is the process towards transforming waste into a reusable material or item. Recycling includes breaking down the materials and utilizing them to make another item that is frequently of lesser quality. Upcycling, then again, is a very specific type of recycling that transforms waste into a material or item that is of better quality. It may generate a creative artistic value for the material and a new life [34]. Pre-consumer silk waste from handloom industries in India is gathered by the small traders for manufacturing various silk embellishments and items like bangles, scarfs, silk embroidery yarn and so on [35].

In India, upcycling is a very common and typical practice. Numerous Indian ladies convert handloom banarasi and kanjeevaram silk sarees to salwar kameez, caftans and so on. Conventional Indian tie and dye or printed clothing like bandhani, leheriyas and old cotton ikats are utilized to make leisure wear, kimonos and night outfits [36, 37].

- **Reduce**

The idea of minimising what is manufactured and what is consumed is crucial for the waste hierarchy. The reason behind it is simple to comprehend—if there is less waste generation, at that point there is less to recycle or reuse. The process of reducing starts with an assessment of what you are utilizing and what it is utilized for. By reducing waste, we avoid the unnecessary use of resources such as materials, energy and water [38].

The textile industry uses large quantities of energy which is generated by electricity, fuels, wood, etc. Hand loom weaving is an intricate process. The process of matching designs, checking the weave pattern, etc., require the weaving area to be lit. Weavers use filament lamps or fluorescent tubes both of which required a lot of electricity. Heat is also required for various handloom processing operations such as drying of cloth, dyeing, scouring, soaping, etc., which is produced by burning of fuels and wood. Consumption of non-renewables sources of energy can be reduced by the use of renewable sources such as solar energy. Solar water heater and solar air-drying systems can be used in handloom processing. Also, conventional LED lamps can be replaced with solar lamps.

The traditional hand loom, with its challenging combined hand–leg movement, is a reason for stress and fatigue to weavers. Also, weavers are disincentivized to keep rehearsing this generations old art specialty given competition from power loom machines and huge textile mills, imported cotton products, increased cost of yarn and inadequate supply of great quality raw cotton, all of which have minimized their profit throughout the long term. On account of familiarity with sustainability researchers have created solar energy-based power looms. This loom consolidates the four limb movements of a traditional loom into a solitary activity, improving the profitability of weavers, which thus expands their income while minimizing the drudgery of weaving [32, 39–41].

5.2 *Use of Sustainable Fibres*

In the textile sector, sustainability begins from agriculture and move towards the best approach to retail. There are numerous things associated with agribusiness industrialization that have serious effect on the environment on the grounds that a ton of pesticides, synthetic fertilizers and water needed in agriculture. Because of this surface run-off of nitrogen, phosphoric emissions of nitrogen compound and leakages have negative affect on the environment. Things continued to change with the time and textile scientists attempted to discover answer for the effects of textile industry on the environment. In this way, an ever-increasing number of textile producers, manufacturers attempted to utilize and search for biodegradable and sustainable fibres for their products to overcome the effect of toxic chemical substances and non-eco processes. They have also found some solutions by new and innovative technologies and processes. Natural fibres like organic cotton, pineapple, bamboo, flax, hemp, jute, ramie, sisal, abaca, etc., are the examples of sustainable fibres in textile industry [42]. Few examples of application of sustainable fibres in handloom sector are given below:

- **Lotus fibre**

Lotus is considered as a very spiritual plant and the motif of lotus flower is a very popular design in textiles. In olden days, Cambodian monks were believed to wear natural-dyed lotus fabric as a symbol of peaceful living, purity and divinity. To extract fibres from stems of the lotus plants are gathered, cut, snapped and twisted to expose their strands. These are thin and white filaments around 20–30 in number, which are rolled into a solitary thread. Yarns are made by arranging the filaments on a bamboo spinning frame and transferring the thread into winders for warping. With much consideration, not to get tangles, threads up to 40 m long are made. These strings are then taken from the warping posts and are curled into big plastic packs. Yarns for the weft are wound into bamboo bobbins. Yarns are woven in manual weaving machines. During the weaving process, yarns are regularly saturated with water by sprays, as the lotus fibres should be kept cool. The fibres are exceptionally delicate and ought to be woven inside 24 h of being extracted in order to forestall their deterioration. The fabric resembles a mix of linen and silk and is resistant to wrinkles and possesses breathable properties given by the molecular makeup of the lotus plants. Lotus fabric is commonly used to make robes for high-ranking monks. Also, few brands are selling shirts, scarfs, etc., made from hand-woven lotus fabric [43].

- **Organic fibres**

Organic cotton and organic silk are widely used for manufacturing handloom sarees. The natural organic silk production is more environmentally friendly, peaceful and sustainable act of silk cultivation. For organic silk, the process is the same as with conventional silk. With the only difference that the mulberry trees are grown organically, without any usage of chemicals and raising the silkworms humanely and without the utilization of hormones. However, there seems the ethical issue that to deliver high evaluation raw silk the pupae in the cocoons ought to be killed by boiling them in water. Consequently, there is another type of silk known as peace silk which is which is produced from cocoons, without executing the pupae [44]. Hand-spun silk yarn having fancy appearance can be interweaved with natural and man-made fibres on handloom to produce designer fabrics at more reasonable prices. This fabric can be used to prepare dress materials, shirting and furnishings [45].

Organic cotton is grown without inorganic fertilizers, fungicides, herbicides, insecticides, etc. Natural organic cotton production is organically based as opposed to chemically dependent on developing frameworks for cultivating. For the development of natural organic cotton, all inorganic fertilizers are precluded and supplanted by farmyard manure, composite, green manure excrements, fish meal, cotton seed meal, cowhide meal, cake, gypsum and so on Organic pesticides, for example, neem cake, ipomea, etc., and natural herbicides are utilized for organic cotton cultivation rather than chemical substance pesticides, insecticide sprays and herbicides for conventional organic cotton production. Likewise, chemical defoliants are denied inorganic harvesting. Organic cotton isn't just significant in the clothing chain but also in the food chain. Whereas with conventionally developed cotton, the pesticides deposits from the cotton seeds concentrate in the fatty tissues of animals and end up in meat and dairy items. Natural organic cotton production additionally helps in minimizing the expense of cotton cultivation by eradicating the utilization of different agrochemicals [46]. Hand-woven organic fabric is widely used for women's apparels such as sarees, dresses, tops, bottoms, blouses, jackets, overlays and throws. The several brands which deal with hand woven organic clothing are ethicus, bhusattva, eleven, doodlage, etc. [47, 48].

- **Banana fibre**

The banana plant (musa sapietuum) belongs to the monocotyledon family. It is cultivated in the tropical regions. After the fruit production, the trunk of the banana plant, i.e. the pseudostem is thrown as farming waste by a large extent. The sheath of the stem is usually scratched by a blunt blade or a decorticator to get the fibre [49]. Raw fibre is cleaned by utilizing soluble base like sodium hydroxide to eliminate pectin and other surface impurities. Each strand of the fibre is taken out, softened with chemicals and woven on handloom after being dyed with different colours. Banana fibres can be blended with other natural fibres like cotton, silk. Hand-woven banana fibre sarees and shirts have great demand in Indian market [49]. Other applications of banana fibres include manufacturing ropes, strings, cables, cords and ship building thread. It is also utilized to manufacture sacks, packing fabrics as well as rugs, mats, high-quality security/currency paper, packing cloth for agriculture products, wet drilling cables, etc. [51]

- **Pineapple fibre**
  Pineapple is perennial herbaceous plant with a height of 1–2 m and width. It usually belongs to the family bromeliaceae. It is predominantly cultivated in tropical regions and coastal regions, majorly for its fruit purpose. There are different strategies to extract the fibrous strands from the leaves of pineapple. Few of them are mentioned below:

  - **Scrapping method of extraction**
    Scrapping machine is the machine utilized for scrapping the pineapple leaf fibre. The machine is the combination of three rollers: (a) feed roller, (b) leaf scratching roller and (c) serrated roller. Feed roller is utilized for the feeding of leaves into the machine; at that point, leaves are sent to the second roller that is called scratching roller. It scratches upper layer of leaf and eliminates the waxy layer. Finally, leaves go to the thick appended sharp edge serrated roller, which pounds leaves and makes a several breaks for the entry passage for the retting organisms.

- **Retting method of extraction**

  In retting process, small bundles of scratched pineapple leaves are inundated in a water tank. Urea or diammonium phosphate is added for fast retting. Pineapple leaves in water tank are consistently checked by utilizing finger to guarantee fibre are loosened and can remove numerous chemical constituents like lignin, fat and wax, ash content, pectin and nitrogenous matter. After completion of retting process, fibre filaments are isolated mechanically, through continuous washing in lake water. Extracted fibres are dried in air by hanging method [52].
  Since pineapple fabric is hand loomed by very few weavers across the world, it is very prized and rare, which is another reason for making it expensive. It is the finest of all Philippine hand-woven fabrics. The most important end use of pineapple fibre is the barong tagalong, wedding dresses and other traditional Philippine dresses. It is also used for table linens, bags, mats and other clothing items. In India, handloom sarees made from pineapple fibres are very popular. Pineapple fibre is often blended with cotton, silk, abaca and linen to create wonderful light, breezy fabrics [53–55].

5.3 Sustainability of the Handloom Sector

To accomplish sustainability in the handloom industry, it is important to present new models for supply of raw material; at production level, intercessions are needed in bringing new plans, designs, patterns, dyeing techniques, information technology should be presented for designs, accounting and marketing; distinctive credit models are needed to executed. Sensitization of banks in handloom concentration areas on topics such as sanctioning weaver loans, proper assessment of working capital requirement, etc. While State Cooperative Banks would have to take a lead in this, Commercial Banks should also be encouraged to provide working capital finance to good working societies. Effective training to the weavers in use of new methods of marketing handloom product should be imparted. This may involve greater use of local haat bazaar, use of mobile sales points, greater interactions between the buyers and sellers, increased use of web marketing, etc. All these are new areas and region-specific suitable models would need to be evolved.

5.4 Sustainable Wet Processing in Handloom Sector

Several eco-friendly approaches for wet processing are mentioned as follows.

- **Effluent treatment plant for handloom units**

The process that is followed in handloom industries is spinning of fibre to yarn, then sizing is applied to improve stiffness of the yarn, later desizing and scouring is carried out to remove excess sizing materials, along with pectin and wax from the yarn/fabric, further bleaching is implemented to remove natural colour from yarn and dyeing generates voluminous quantities of effluent and in turn causes environmental pollution. The effluents consist of high concentrations of the dye stuff, biochemical oxygen demand and dissolved solids. In the recent years, water quality in metropolitan regions and villages adjacent to dyeing industrial areas has depreciated attributable to dye effluent inflow into land and water bodies. Unlike industrial sectors that go under the domain of the diverse pollution control measures, small-scale handloom units don't go under these measures. Subsequently, there is a huge involvement of toxic pollutants because of the release of effluents into nearby water bodies and ground water pollution. In this manner, it is exceptionally vital to treat the dye effluent prior to releasing into the environment [56].

Sri Lankan handloom industry sets good example of sustainable wet processing. They treat wastewater released from handloom dyeing plants to meet environmental standards, then treated water is used for agricultural purposes [57]. In another study, use of solar energy to treat handloom waste water is recommended. In the process, solar dryer has used which converts solar radiation to heat. This heat is used to evaporate waste water from handloom dyeing units and the remaining dye will be alone deposited on the bottom surface. This dye will be further used for handloom units [58].

- **Use of enzymes**

Biotechnology offers the potential for new industrial processes that require less energy and depend on inexhaustible raw materials. The utilization of enzymes in processing of textiles is already established industrial technology. Treatment of textiles with enzymes, such as cellulosic materials, for example, cotton, gooey or lyocell textures, have generally been utilized in the textile process houses since the 1980s. Enzymatic handling empowers the textile business to minimize production costs, decrease the ecological effect of the overall process and to improve the superiority and usefulness of the final products. Proteins are non-poisonous and they require gentle temperature and pH. Today enzymatic treatment of cotton either in denim washing or in bio stoning is standard method used in industry. Handloom cotton has a few inadequacies, as higher maintenance costs for washing and iron pressing, partly for rough texture and less durability. Cellulase enzymes are exceptionally viable in eliminating loose fibres from the textile surface, a process known as biopolishing. An attempt is made to treat the handloom cotton fabric using cellulase enzyme in order to improve softness and smoothness [59].

Natural silk comprises of two proteins, specifically sericin and fibroin, which vary impressively in their chemical composition and availability. The degumming cycle might be considered essentially as an interaction of cleavage of peptide bonds of sericin by the hydrolytic or enzymatic strategy and its resulting expulsion from fibroin by solubilization or scattering in water. Degumming of silk has been generally completed with alkali and soap. These techniques have some significant drawbacks, for example, the degummed silk got isn't uniform in quality, the strength loss is high and the chemicals utilized can cause environmental pollution. Enzymes are now being considered as alternative degumming agents for the processing of silk. Authors have reported the use of microbial protease for degumming of handloom silk fabric [60, 61].

- **Water conservation**

Water utilization might be diminished by receiving most recent colouring methods and empowering reuse of dye liquor for dyeing. Most recent synthetic substances, apparatus which dye at lower liquor proportions ought to be utilized to lessen pollution.

- **Use of natural dyes**

With expanding attention to the ecological contamination and health hazards related with synthesis, handling and utilization of synthetic colours has made the preservationists raise the call 'return to the nature', if the world is to be saved from moderate harming. Interest in normal product is developing all through the world and individuals are getting mindful of the requirement for eco-friendly materials to come up and overwhelm the scene [62].

Synthetic dye substances cause allergies in human beings and also have several carcinogenic properties. On the other hand, since natural dyes are obtained from renewable resources such as plant, animal, mineral and microbial and they do not have any health hazards. They are biodegradable and are non-toxic. In addition to that, some of the natural dyes have good antioxidant and medicinal properties. Because of which they are commonly used in the cosmetic, food and textile industries [63].

Several attempts were made by researchers for application of natural dyes on handloom fabric. For instance, batik work with natural dyes extracted from plants was carried out on handloom cotton fabric [64]. In another study, painting on handloom cotton fabric was carried out using plant extract in presence of mordant [3]. In India, traditionally most of the handloom sarees like paithani, gadwal, chanderi, patola, etc., are dyed with natural dyes extracted from plants. The two trades of handloom weaving and natural dyeing were inseparably interwoven. Natural colour improves the allure of the hand-woven fabric complex; nay, without colouring, the varieties of the designs in the woven material may not be clear and their uniqueness might be completely lost [65].

In another research, endeavours were made to make hand-woven khadi textured fabrics more appealing with creative designs through printing with colourants removed from natural resources in presence of different eco-friendly mordants [66].

6 Conclusion

This chapter discussed the positive and negative aspects of environmental sustainability of handloom sector. The main aim is to achieve sustainability in handloom sector. Handloom manufacturing involves a lot of mechanical and chemical processing which consumes a lot of energy, chemicals and dyes. Often Natural dyes are preferred over synthetic dyes to decrease pollution. Natural dyes are eco-friendly and are offer many advantages over synthetic dyes, particularly in enhancing and maintaining health. These dyes have natural healing and medicinal properties. Further, energy consumed is also generally derived from thermal power plants and therefore it generates large amount of greenhouse gases which have a negative impact on the environment. Instead renewable energy sources such as solar and wind energy should be utilized which will reduce the power cost and reduce generation of greenhouse gases and carbon emission. Also, handloom spinning generates a large amount of solid waste, such as dust, lint, creates noise pollution and air pollution. Handloom industries are facing several challenges due to the increased cost of raw materials, increased labour wages and increased sustainability driven global fashion sectors. This chapter also discusses various approaches in handloom sector to achieve sustainability by implementing the 3Rs (Reduce, Recycle and Reuse) in the waste management system. By adopting many of the approaches mentioned in this chapter, the handloom sector can become more sustainable in India and developing countries.

References

1. Broudy E (1993) The book of looms: a history of the handloom from ancient times to the present. UPNE
2. Murray K (2011) Sustainability in craft and design. Craft+ Design Enquiry, vol 3, p 1
3. Maulik SR, Agarwal K (2014) Painting on handloom cotton fabric with colourants extracted from natural sources. Indian J Trad Knowl 133):589–595
4. Dissanayake DGK, Perera S, Wanniarachchi T (2017) Sustainable and ethical manufacturing: a case study from handloom industry. Text Cloth Sustain 3(1):1–10
5. Soundarapandian M (2002) Growth and prospects of handloom sector in India. National Bank for Agriculture and Rural Development
6. Bhalla K, Kumar T, Rangaswamy J (2018) An integrated rural development model based on comprehensive Life-Cycle Assessment (LCA) of Khadi-Handloom Industry in rural India. Procedia CIRP 69:493–498
7. Mishra V, Bhattacharjee M (2017) Sustainability of Handloom Value Chain-A Case Study of Nadia District in West Bengal. Int J Econ Res 14(3):277–288
8. Rashid BHT, Yousaf I, Rasheed Z, Ali Q, Javed F, Husnain T (2016) Roadmap to sustainable cotton production. Life Sci J 13(11):41–48
9. Karthik T, Rathinamoorthy R (2017) Sustainable silk production. In: Sustainable fibres and textiles. Woodhead Publishing, pp 135–170
10. Sivaram NM, Gopal PM, Barik D (2019) Toxic waste from textile industries. In: Energy from toxic organic waste for heat and power generation. Woodhead Publishing, pp 43–54
11. Mia R, Selim M, Shamim AM, Chowdhury M, Sultana S (2019) Review on various types of pollution problem in textile dyeing & printing industries of Bangladesh and recommandation for mitigation. J Text Eng Fashion Technol 5(4):220–226

12. Nahar K, Chowdhury MAK, Chowdhury MAH, Rahman A, Mohiuddin KM (2018) Heavy metals in handloom-dyeing effluents and their biosorption by agricultural byproducts. Environ Sci Pollut Res 25(8):7954–7967
13. Kumar S (2018) Impact of low illumination and high noise level on occupational health of Indian handloom weavers: Special reference to Bargarh district of Odisha (Doctoral dissertation)
14. Alexander TH, Sakshi GA, Manisha G (2020) Quality characteristics of recycled handloom fabrics. Pharma Innovat J 9(3):350–352
15. Muthu SS, Li Y, Hu JY, Ze L (2012) Carbon footprint reduction in the textile process chain: recycling of textile materials. Fibers Polymers 13(8):1065–1070
16. Mensah J, Casadevall SR (2019) Sustainable development: Meaning, history, principles, pillars, and implications for human action: Literature review. Cogent Soc Sci 5(1):1653531, 1–21
17. Barooah N, Dedhia EM (2015) Study of socio-economic status of women engaged in handloom weaving and measures for enhancing their sustainability. Int J Res Soc Sci 5(4):653–665
18. Alhaddi H (2015) Triple bottom line and sustainability: A literature review. Bus Manage Stud 1(2):6–10
19. Aleksandra Kokić Arsić, Milan Mišić,Mladen Radojković, Bojan Prlinčević. In: 1st international cconference on quality of life, pp 83–88
20. Mamidipudi A, Bijker W (2012) Mobilising Discourses: handloom as Sustainable socio-technology. Econ Politic Weekly, pp 41–51
21. An introduction to environmental degradation: Causes, consequence and mitigation Pradip Kumar Maurya, Sk Ajim Ali Ateeque Ahmad, Qiaoqiao Zhou, Jonatas da Silva Castro4Ezzat Khan, Hazrat Ali (2020). Environmental degradation: Causes and Remediation Strategies. Vol. 1). Agro Environ Media, Publication Cell of AESA, Agriculture and Environmental Science Academy, pp 1–21
22. R Chopra 2016 Environmental degradation in India: causes and consequences. Int J Appl Environ Sci 11(6):1593–1601
23. Choudhary MP, Chauhan GS, Kushwah YK (2015) Environmental degradation: causes, impacts and mitigation. In: National seminar on recent advancements in protection of environment and its management issues (NSRAPEM-2015).
24. R Goodland 1995 The concept of environmental sustainability Annu Rev Ecol Syst 26(1):1–24
25. Nilofar Nisha J, Arun Prakash M, Vignesh P, Bharath Ponvel M, Kirubakaran V, Renewable energy integrated waste water treatment for handloom dying units: an experimental study
26. G Foy 1990 Economic sustainability and the preservation of environmental assets. Environ Manage 14(6):771–778
27. Balaji NC, Mani M (2014) Sustainability in traditional handlooms. Environ Eng Manag J (EEMJ) 13(2)
28. Prasad NRIG, Belli ISKM (2014) Contemporary issues and trends in fashion. retail and management. In: International conference on fashion, retail and management. NIFT
29. Gupta V (2012) Recycling of textile waste in small clusters and its contribution to the socioeconomic upliftment of the community. AsiaInCH Encyclopedia India
30. Yalcin-Enis I, Kucukali-Ozturk M, Sezgin H (2019) Risks and management of textile waste. In: Nanoscience and biotechnology for environmental applications. Springer, Cham, pp 29–53
31. Fortuna LM, Diyamandoglu V (2017) Optimization of greenhouse gas emissions in second-hand consumer product recovery through reuse platforms. Waste Manage 66:178–189
32. Kamis A, Suhairom N, Jamaluddin R, Syamwil R, Puad FNA (2018) Environmentally sustainable apparel: recycle, repairing and reuse apparel. Int J Soc Sci Human Invent 5(1)4250–4257
33. Filho L, Dawn Ellams W, Han S, Tyler D, Boiten VJ, Paço A, Moora H, Balogun A-L (2019) "A review of the socio-economic advantages of textile recycling." J Cleaner Prod 218:10–20
34. Hassan A (2020) Upcycling versus recycling-a way to protect the environment during COVID-19 lockdown
35. Bairagi N (2014) Recycling of textiles in India. J Textile Sci Eng 3(1):4
36. Singh J, Sung K, Cooper T, West K, Mont O (2019) Challenges and opportunities for scaling up upcycling businesses–The case of textile and wood upcycling businesses in the UK. Resour Conservat Recycling 150:104439

37. Sumod P, Mishra K, Rangnekar S, India: the catalyst for a sustainable future by rediscovering recycling habits
38. Jadhav NC, Jadhav AC (2020) Waste and 3R's in footwear and leather sectors. In: Leather and footwear sustainability. Springer, Singapore, pp 261–293
39. Gupta S (1989) Scope for solar energy utilization in the Indian textile industry. Sol Energy 42(4):311–318
40. Hasanbeigi A (2010) Energy-efficiency improvement opportunities for the textile industry (No. LBNL-3970E). Lawrence Berkeley National Lab.(LBNL), Berkeley, CA (United States), pp 110–112
41. Harnetty P (1991) Deindustrialization Revisited: The Handloom Weavers of the Central Provinces of India, c. 1800–1947. Modern Asian Stud 25(3):455–510
42. Blackburn R (ed)(2009) Sustainable textiles: life cycle and environmental impact. Elsevier
43. Aishwariya S, Thamima S (2019) Sustainable textiles from lotus. Asian Textile 28(10):56–59
44. Sannapapamma KJ, Naik SD (2015) Ahimsa silk union fabrics–A novel enterprise for handloom sector. Indian J Trad Knowl 14(3):488–492
45. Prof. Dr. Panomir Tzenov, Asoc. Prof. Dr. Diliana Mitova, Dr. Eng. Maria Ichim, The Organic Sericulture in the Context of Biological Agriculture and Organic Textile, 7th BACSA INTERNATIONAL CONFERENCE "Organic Sericulture – Now and the Future" "ORGASERI" 2015 Sinaia, Romania April 19th–24th 2015, pp 6–18
46. Shaikh MA (2011) Sustainable eco-friendly organic cotton. Middle East 52:49–450
47. Sharda NL, Kumar VKM (2012) Multifarious approaches to attain sustainable fashion. Nordic Text J 1:31-37
48. Khandual A, Pradhan S (2019) Fashion brands and consumers approach towards sustainable fashion. In: Fast fashion, fashion brands and sustainable consumption. Springer, Singapore, pp 37–54
49. Sinha MK (1974) 5—The use of banana-plant fibre as a substitute for jute. J Text Inst 65(1):27–33
50. Kordhanyamath J, Bai SK (2019) Speciality of Banana Yarn on Ilkal Handloom Sarees Woven with Murgi Motif
51. Neelam S, Bharti S (2017) Banana: eco-friendly fibre used for household articles. Asian J Home Sci 12(2):642–646
52. Banik S, Nag D, Debnath S (2011) Utilization of pineapple leaf agro-waste for extraction of fibre and the residual biomass for vermicomposting. Indian J Fibre Text Res 36:172–177
53. Das PK, Nag D, Debnath S, Nayak LK (2010) Machinery for extraction and traditional spinning of plant fibres
54. Jose S, Salim R, Ammayappan L (2016) An overview on production, properties, and value addition of pineapple leaf fibers (PALF). J Nat Fibers 13(3):362–373
55. Kannojiya R, Gaurav K, Ranjan R, Tiyer NK, Pandey KM (2013) Extraction of pineapple fibres for making commercial products. J Environ Res Dev 7(4):1385
56. Goswami R, Jain R (2014) Strategy for sustainable development of handloom industry. Global J Finance Manag 6(2):93–98
57. Wanniarachchi T, Dissanayake K, Downs C (2020) Improving sustainability and encouraging innovation in traditional craft sectors: the case of the Sri Lankan handloom industry. Res J Text Apparel 1–24
58. Nilofar Nisha, J., Arun Prakash, M., Vignesh, P., Bharath Ponvel, M., & Kirubakaran, V. Renewable Energy Integrated Waste Water Treatment For Handloom Dyeing Units: An Experimental Study. Research Journal of Chemistry and Environment, 24, pp.66–69
59. Mojsov K (2016) Effects of enzymatic treatment on the physical properties of handloom cotton fabrics. Tekstilna industrija 63(1):21–26
60. Gulrajani ML, Agarwal R, Chand S (2000) Degumming of silk with a fungal protease. Indian J Fibre Text Res 25:138–142
61. Freddi G, Mossotti R (2003) Innocenti 2003 Degumming of silk fabric with several proteases. J Biotechnol 106(1):101–112
62. Kar A, Borthakur SK (2008) Dye yielding plants of Assam for dyeing handloom textile products

63. Pandit P, Shrivastava S, Maulik SR, Singha K, Kumar L (2020) Challenges and Opportunities of Waste in Handloom Textiles. In: Recycling from waste in fashion and textiles: a sustainable and circular economic approach, pp 123–149
64. Maulik SR, Bhowmik L, Agarwal K (2014) Batik on handloom cotton fabric with natural dye
65. Ganesh S (2008) The role and development of vegetable dyes in Indian handlooms. Indian J Trad Knowl 7(1):125–129
66. Banerjee AN, Pandit P, Maulik SR (2019) Eco-friendly approaches to rejuvenate the Khadi udyog in Assam. Indian J Trad Knowl 18(2):346–350

Sustainability, Culture and Handloom Product Diversities with Indian Perspective

Sanjoy Debnath

Abstract One of the ancient technological interventions is handloom. With the passage of time, the handloom converted into power loom and gradually shuttleless looms. This handloom is coming under traditional knowledge, and generation after generation this knowledge has been transferred. These encompasses handloom preparatory, handloom operation and product making out of this system. India is one of the traditionally rich countries where a wide diversified product from handloom are being manufactured and sold commercially. Overall, the handloom products are unique in nature and fetch good attraction to niche market. Sustainability is one of the important issues in handloom products and the industry. This chapter focusses different issues of environmental aspects at every step involved in handloom preparatory, handloom weaving and post-operation of handloom fabrics and products. In this chapter, an overview of sustainability of Indian handloom industry, its products and cultural interventions has been covered elaborately.

1. Cultural Intervention and Handloom Product Diversities in India.
2. Handloom Weavers, Their products and Culture, Environmental aspects—Sustainability.
3. Conclusions and Recommendations.

Keywords Diversification · Environmental aspects · Handloom products of India · Indian cultural intervention in handloom products · Indian handloom · Sustainability

S. Debnath (✉)
ICAR-National Institute of Natural Fibre Engineering & Technology, 12, Regent Park, Kolkata, West Bengal 700040, India
e-mail: sanjoy.debnath@icar.gov.in

M. Á. Gardetti et al. (eds.), *Handloom Sustainability and Culture*, Sustainable Textiles: Production, Processing, Manufacturing & Chemistry,
https://doi.org/10.1007/978-981-16-5665-1_2

1 Introduction: Cultural Intervention and Handloom Product Diversities in India

One of the ancient technological interventions is handloom. With the passage of time, the handloom is converted into power loom and gradually shuttle-less looms. This handloom is coming under traditional knowledge, and generation after generation this knowledge has been transferred. In context of India, handloom is only means of income of the family in handloom clusters. Almost all the family members irrespective of age and gender are involved in weaving and weaving preparatory aspects to fulfill the requirements of handloom feed material and handloom product making. As per as the history of handloom [10, 11] is concerned, it starts from carpet during the ancient times, a portion of Indian handloom was excavated from the parts of Egypt. Later on, fine woven-dyed cotton fabrics were found from excavations of Indus Valley Civilization, i.e. Mohenjo Daro. Apart from these, there exists many such historical evidences of ancient Indian handloom from excavations. Subsequent Aryan settlers in the region also adopted and further improved techniques of weaving using cotton wool followed by embellishing these fabrics with dyes and embroidery. These indicate the golden history of Indian handloom and its popularity even in ancient days. In fact, traditional handloom style has been one of the oldest forms as found from the document [11]. Even the Vedic literature also has mention of Indian weaving styles. Apart from this, few examples are also seen in Buddhist era scripts about the woollen carpets.

Prior to imperialism and colonization, all the natural fabrics (silk, cotton, wool and jute) were hand-woven and Khadi was among the prevalent materials at that point of time. Later on, the mechanical system created ways for the faster completion of spinning and weaving. Furthermore, with the development of science and technologies, machineries were introduced, the technicalities were given due care and the finesse in product came in a better way. This also helped the weavers, embroiders and hand-printers to create new designs and product diversities. At the time of British, the export business of cotton and silk also started. This enabled the Indians to showcase the talent of their expertise in other countries also. In context to India, following are the products branded under handloom products [9]:

- Sarees.
- Dress material/fabrics.
- Shawls.
- Stole/dupatta.
- Scarf.
- Muffler.
- Home furnishing.
- Bed linen.
- Kitchen linen.
- Table linen.
- Garments of Indian handloom-based fabrics.
- Hand bag from handloom fabric.

- Warm clothings from handloom fabric.
- Rugs and carpet.

Indian handloom created a special place for itself in India as well as abroad. With so many varieties of handloom from different states, India has collected a precious wealth of innovation. After all, this has led to the emergence of India as the most richly cultured country. These encompass handloom preparatory, handloom operation and product making out of this system. India is one of the traditionally rich countries where a wide diversified product from handloom is being manufactured and sold commercially. Overall, the handloom products are unique in nature and fetch good attraction to niche market. Sustainability is one of the important issues in handloom products and the industry. This chapter focusses different issues of environmental aspects at every steps involved in handloom preparatory, handloom weaving and post-operation of handloom fabrics and products. In this chapter, an overview of sustainability of Indian handloom industry, its products and cultural interventions has been covered elaborately.

As per as the definition of 'Handloom' is concerned, it can be defined as 'handloom' is a loom that is used to weave cloth without the use of any electricity [12]. Hand weaving is done on pit looms or frame looms generally located in weavers' homes. Weaving is primarily the interlacing of two sets of yarn—the warp (lengthwise) and the weft (widthwise). Mostly, handlooms in India are found in clusters at different parts of the country with product diversities. Government of India under Ministry of Textiles, promoting 'Handloom Brand' for traditional hand-woven heritage of India and also assuring quality product to the consumer. India has a long tradition of excellence in making high-quality handloom products with extraordinary skills and craftsmanship, which are unparalleled in the world.

Since the ancient times handlooms are used as per the suitability in different types of handlooms as follows [11]:

Ancient Looms: The first and original loom was vertically twist-weighted types, where threads are hung from a wooden piece or branch or affixed to the floor or ground. The weft threads are manually shoved into position or pushed through a rod that also becomes the shuttle. Raising and lowering each warp thread one by one is needed in the beginning. It is done by inserting a piece of rod to create a shack, the gap between warp threads in order for the woof to easily traverse the whole warp right away.

Ground Looms: Horizontal ground looms permit the warp threads to be chained between a couple of rows of dowels. The weaver needs to bend forward to perform the task easily. Thus, pit looms with warp chained over a ditch are invented to let the weaver have his or her legs positioned below and levelled with the loom.

Back-strap Looms: They are well recognized for their portability. The one end of this loom type is secured around the waist of the weaver and the other end is attached around a fixed thing like door, stake or tree. Pressure applied can be customized by just bending back.

Frame Looms: Frame looms almost have the similar mechanisms that ground looms hold. The loom was made of rods and panels fastened at the right angles to construct a form similar to a box to make it more handy and manageable. This type of loom is being utilized even until now due to its economy and portability.

Rigid heddle Looms: These are the crisscross manifold loom types. The back-strap looms and frame looms fall under this type. This one normally features one harness, with its heddles attached in the harness. The yarn or thread goes in an alternate manner all the way through a heddle and in the gap between the heddles. In this way, lifting the harness also lifts half of the threads and letting down the harness also drops the same threads. Strands leading through the gaps between the heddles stay in position.

Foot-treadle Floor Looms: Nowadays, hand weavers are likely to employ looms having no less than four harnesses. With every harness featuring a set of heddles wherein wool can be strung, and by lifting the harnesses in diverse arrangements, a multiplicity of designs is created. Looms having a couple of harnesses similar to these are applied for knitting tabby, the unvarying weave textiles.

Haute Lisse and Basse Lisse Looms: These are generally employed for knitting conventional tapestry. Haute lisse has the yarn or thread hung straight up between two spools. The basse lisse loom has the warp thread stretched out horizontally between spools.

Shuttle Looms: It is the key component of the loom along with the warp beam, shuttle, harnesses, heddles, reed and take up roll. In the loom, yarn processing includes detaching, battening, alternative and taking-up operations.

2 Handloom Weavers, Their Products and Culture, Environmental Aspects—Sustainability

In India, handloom weavers are practicing for making products from generation to generation and mostly the cultural reflection found in their products. India is very rich in handloom product diversities and its handloom products, one can easily classify the geographical location and the motif/design which are also unique in nature. It has been documented by [2] details of Indian traditional handloom products emphasizing saree from location-specific products discussed below [9].

Ikat from Odisha

Ikat is weaved in several parts of the country but Odisha seems to have pretty much mastered the Ikat art of weaving. Ikat weavers in Odisha are more often than not the members of communities like the Meher or Bhulia who have inherited the art form and have mastered the trait over the years. They basically strive to bring the rich Oriya culture to life in their Ikats with their unique dyeing techniques. And today as Ikat makes a raving comeback to the ramps with contemporary design labels like Vraj:Bhoomi and The Ikat Story, we cannot take away from the weavers their due in making the fabric so popular (Fig. 1).

Fig. 1 Yarn dyed plain woven designed fabric (*Ikat*)

Kalamkari from Andhra Pradesh

This is one of the most prominent features of Andhra Pradesh, Kalamkari is a kind of hand-painted or block-painted textile art that dates back to the Indus Valley Civilization. With a history of 3000 years, Kalamkari is known to have evolved during the Mughal era and has managed to retain its grace till date. Today as we watch Kalamkari make a contemporary fashion statement with conventional designers, we think we owe it to the skilled artisans of Venkatagiri, Pochampali and Gadwal that continue to create traditional and contemporary pieces adorned with Kalamkari prints (Fig. 2).

Bandhani or tie and dye from Gujarat

A unique design made by tie and dye or *bandhani* is a derivative of the Sanskrit word 'bandh' which means 'to tie' and is one of the most popular textile arts of India. Considering they come from one of the most culturally rich states of Gujarat and Rajasthan, bandhani in its full glory is a burst of vibrant colours and glass work.

Fig. 2 *Kalamkari* designed fabric

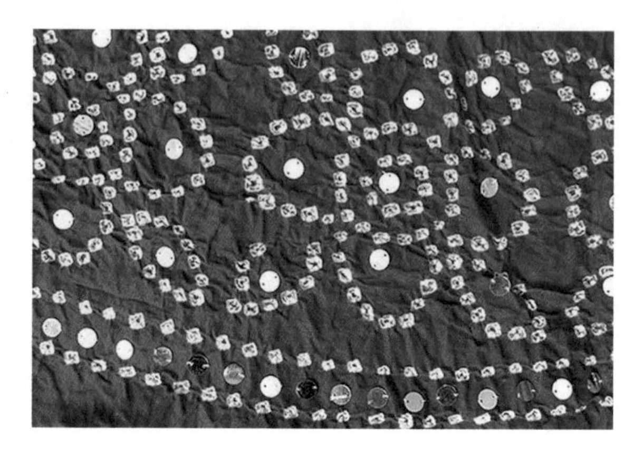

Fig. 3 Tie and dye (*Bandhani*)

The authentic bandhani features square or round motifs that are a result of the dye—the more intricate your tie and dye the more authentic your bandhani. Like most traditional fabrics, while machinery might have increased production, true tie and dye remains native to the craftsmen and their skill alone (Fig. 3).

Patan Patola from Gujarat

This product is made in pure silk, Patola sarees are the ultimate manifestation of the weaving prowess that the artisans of Gujrat have attained over several centuries. In the Patan Town of Gujarat, this silk cloth with double Ikat patterns is brought to life as weavers work diligently for over 5 months to weave one Patola saree. The intricate weaving and dying techniques add to the authentic exoticism of the fabric (Fig. 4).

Brocades from Uttar Pradesh

Banarasi sarees are one of India's most precious textile art forms and wearing one is like wearing a piece of art. The weavers of Varanasi are nothing short of artists themselves as they weave with fine gold and silver metallic threads to create exotic delicate brocades. This high-quality weaving procedure has not only managed to sustain itself through the years but has also bounced back into limelight for its sheer gorgeousness and beauty (Fig. 5).

Zari work from Madhya Pradesh

The cities of Bhopal, Gwalior and Indore are known for the intricate Zari work which was patronized by the Mughal Emperors over 300 years ago. And while technically Zari is a brocade of tinsel threads meant for weaving and embroidery, once it is woven into the fabric (mostly silk) to create various patterns, it makes the fabric its own (Fig. 6).

Kancheepuram from Tamil Nadu

The roots of Kancheepuram sarees trace back to the town of Kanchi or Kancheepuram of Tamil Nadu and are known for their exquisite pallus that are laden with Zari work. These sarees are so elaborate in grandeur that they take anything between 10 and

Fig. 4 *Weaver weaving Patola* saree in handloom

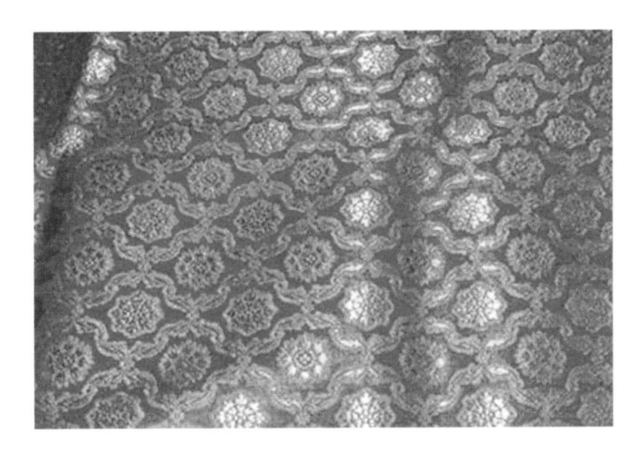

Fig. 5 *Banarasi saree*—a product of Varanasi (Banaras)

20 days, even 6 months at times to be woven in. And because all the fuss is worth it, these Kancheepuram sarees continue to hold their spot in bridal trousseaus even today (Fig. 7).

Balarampuram Sarees of Kerala

It is the traditional saree produced at Blarampuram in Thiruvanathapuram District of Kerala. It is specifically in golden ribs weave design. The design appears identical on both sides of the fabric. There is no rough or unfinished-like appearance on back

Fig. 6 Zari work from Madhya Pradesh

Fig. 7 Kancheepuram saree from Tamil Nadu

side of the fabric. The basic texture, i.e. thread density of the fabric is higher than that of similar cotton sarees with less starchy finishing in the saree providing soft and comfort touch (Fig. 8).

Chendamangalam Dhoti of Kerala

It is a very popular *Dhoti* produced at Chendamangalam in Ernakulam District of Kerala, woven with undyed grey cotton yarn. This design appears very prominently

Fig. 8 *Balarampuram* sarees of Kerala origin

on both side borders and cross border. The basic texture is thread density of the fabric which is higher than that of similar cotton dhotis. This woven dhoti with also less starchy finish provides softer touch (Fig. 9).

Chettinadu Cotton Saree of Tamil Nadu

This is a traditional saree woven at Chettinadu in Shivaganga District of Tamil Nadu which has small strips in different colours at the joining of boarders in warp way on both sides. To achieve this, simple extra warp designs are employed and no extra wefts are used. These sarees are comparatively thicker and have broad stripe or check patterns (Fig. 10).

Kasargode Saree of Kerala

It is a traditional saree woven in Kasargode of Kerela having stripes, check patterns in tie and dye technique to obtain solid border on pallu appearing very prominently in the fabric. Normally, high thread density is used in the fabric compared to similar cotton sarees. There is lesser starchy finish (loom sizing technique) in the saree, so it is so soft to touch (Fig. 11).

Khandua silk saree of Odisha

Khandua silk of ikat produced in various districts of Odisha in different forms of Pochampali and Patola of Gujarat because of yarn grouping systems. Geometrical patterns are the specialty found in Patola or Pochampali sarees; however, continuous floral pattern and shaded effects are seen in Odisha Ikats. Designs on face and back sides of the fabric appear same which is the speciality of the design. Blurriness in design is noticed in the design due to yarn tie-dye technique used (Fig. 12).

Kuthampully Saree of Kerela

This is a traditional saree woven at Kuthampally in Thrissur District of Kerela in which designs appear very prominently in side borders, body and pallu. Sarees are

Fig. 9 Chendamangalam dhoti of Kerala

Fig. 10 Chettinadu cotton saree of Tamil Nadu

Fig. 11 Kasargode saree of Kerala

mostly woven with undyed yarns like Balarampuram sarees. Apart from these, mostly complex designs and motifs are found in the sarees. These sarees are softer in feel due to less starch application (Fig. 13).

Fig. 12 Khandua silk saree of Odisha

Fig. 13 Kuthampully saree of Kerela

Matka noil home furnishing

This product is produced in Bhaghalpur, Bihar, Malda, and West Bengal and fabric is densely woven and used for fabric used for wide range of home furnishing. The surface of the fabric has short fibres which gives rough texture on the fabric. The fabric handle is soft when compared to similar cotton fabric. To achieve such product, different woven structures like basket, hounds-tooth, diamond, etc. are used.

Rib Mat/ Place Mat

This mat is unique and produced in Kerala, Tamil Nadu and Panipat, and has thick ribs in weft or horizontal direction. It is one of the components of table linen. Normally, it is rectangular piece of cloth of size 30–35 cm in width and 45–50 cm in length. It is plain woven structure and firmly woven to give firm texture (Fig. 14).

Runner/Place Mat

It is the product of Kerela, Tamil Nadu and Panipat which used as components of table linen. Size of the fabric is normally in the range of 33–35 cm in width and 150–200 cm in length. It is generally plain woven structure, coarser in texture, striped or checked design depending upon market demand (Fig. 15).

Fig. 14 Rib mat/place mat

Fig. 15 Runner/place mat

Tussar Sikl Saree/Fabric

Hand reeled Tussar (kosa) silk is used in natural colour in weft direction and the product is produced in Chattisgarh, Bihar, West Bengal and Odisha. Rough, coarse texture and weft bars are visible due to unevenness of the yarn. Colour varies from yellowish beige to brown in natural state. Causes crease/wrinkle easily especially in water. The fabric creates rusting sound when rubbed together (Fig. 16).

Baluchari Silk Saree

Traditional saree woven in Bishnupur in West Bengal can easily be identified from the construction of long pallu and placement of its design motifs in perfect rectangular closed corners maintaining continuity of the designs without break. Basic fabric is heavier and more compact than Benerasi saree. Designs are made with extra weft using silk yarn and no zari is used as done in Benerasi or other silk sarees (Fig. 17).

Chanderi Saree/Fabric

Produced at Chanderi in Ashok Nagar District of Madhya Pradesh in which warp thread is un-degummed mulberry silk whereas weft thread is cotton, which is not commonly used in any textile products. Due to un-degummed silk warp, the fabric texture is somehow not very soft. The fabric woven in compact structure and it is transparent and lightweight which is suitable as summer wear (Fig. 18).

Maheshwari Saree/Fabric

This is a traditional product woven at Maheshwar in Khargon District of Madhya Pradesh in which generally the boarder design of the saree is reversible. The design looks similar from both sides. In warp, un-degummed mulberry silk yarn and in weft cotton yarns are used. These woven sarees are heavier and more compact than Chanderi sarees. Normally, there is no designing work with extra threads in the body of the saree (Fig. 19).

Fig. 16 Tussar silk saree/fabric

Fig. 17 Baluchari silk saree

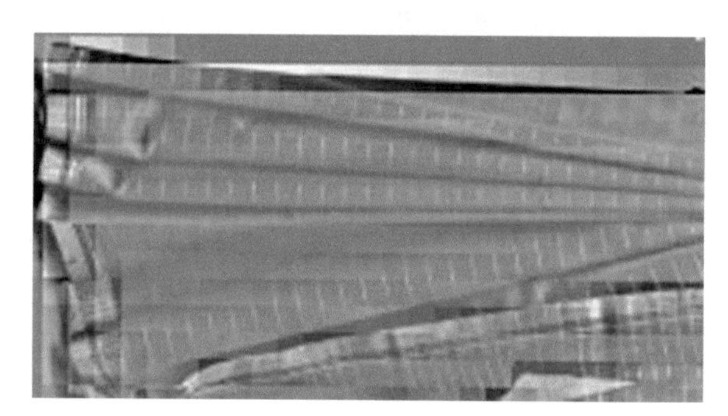

Fig. 18 Chanderi saree

Jamdani Cotton Saree

Jamdani sarees are woven in Nadia and Burdwan Districts of West Bengal. This saree can be distinguished by seeing the extra weft which is usually inserted in the ratio of two ground thread and one design thread. A special bulging effect is seen at the design portions of the fabric since design thread is coarser than ground thread. The extra un-cut weft yarn is interlaced with warp yarns to form the design from left to right and vice versa in such a way that it cannot be pulled off (Fig. 20).

Kullu Stole/Shawl

This is a traditional product of Kullu District of Himachal Pradesh in which designs are developed with different coloured weft threads of short length from Indian wool. No continuous thread is used from one end to other which is used in the design area.

Fig. 19 Maheshwari saree

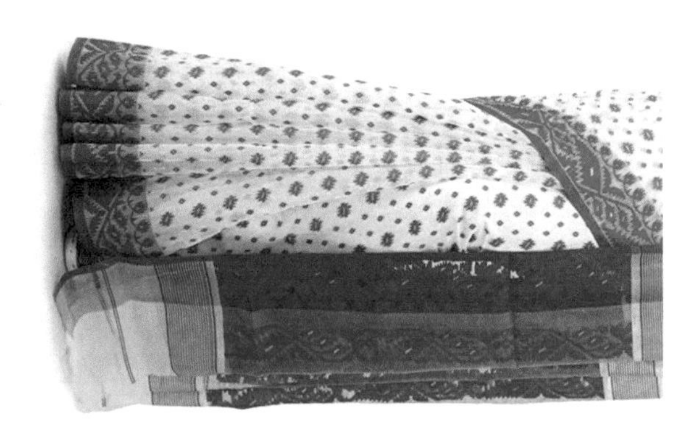

Fig. 20 Jamdani cotton saree

Basic fabric has diagonal twill lines (twill weave design). Kullu shawl is famous for its intricate multi-colour strong geometrical pattern woven with woollen yarns (Fig. 21).

Mangalagiri cotton saree/fabric

Popular product of Mangalagiri in Guntur District of Andhra Pradesh has basic texture more compact than that of similar cotton sarees. Extra warp designs are spread continuously without any gap up to the selvedge of the saree. The texture of the saree is comparatively soft due to no starch finish. Mangalagiri dress materials do not have extra weft designs on the body (Fig. 22).

Pochampali Ikat silk/cotton Saree and cotton bed sheet

This is a traditional product of Pochampali in Yadadri Bhuvanagiri District of Telangana which is perfectly reversable cloth with same appearance of the design on both face and back sides of the fabric. The intensity of the colours in the design also

Fig. 21 Kullu stole/shawl

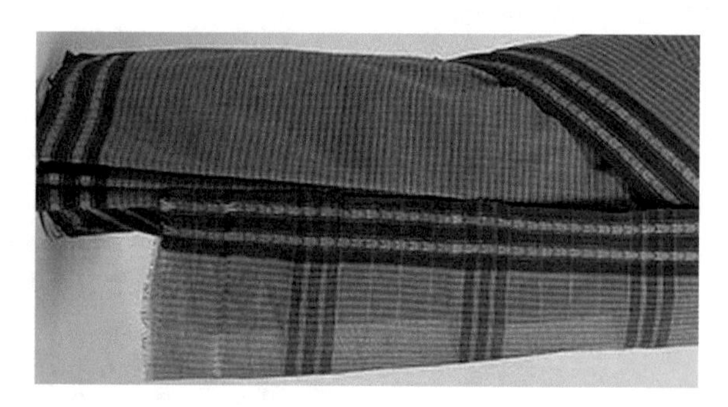

Fig. 22 Mangalagiri cotton saree

appears same on both sides of the fabric, whereas if the fabric is printed, then colours in the back side of the fabric will be lighter. Contours of the design are always hazy. Same appears in case of bed sheet. For all the saree or bed sheets, plain woven design is being used (Fig. 23).

Muga Silk fabric

It is a heritage product of Sualkuchi in Kamrup District of Assam with natural rich golden colour fabric. The texture is glossy and lustrous. Durability is exceptionally good. As the fabric ages its golden lustre increases. The moisture regain of the material is 30% which is much more than other silk fabrics. Muga silk fabric absorbs 85.08% of the harmful ultraviolet rays in sunlight (Fig. 24).

Kota Doria Saree/Fabric

It is a traditional product of Kota and Baran Districts of Rajasthan and basic texture in combination of cotton and silk yarn gives transparent effect on the fabric. One small

Fig. 23 Pochampali Ikat saree

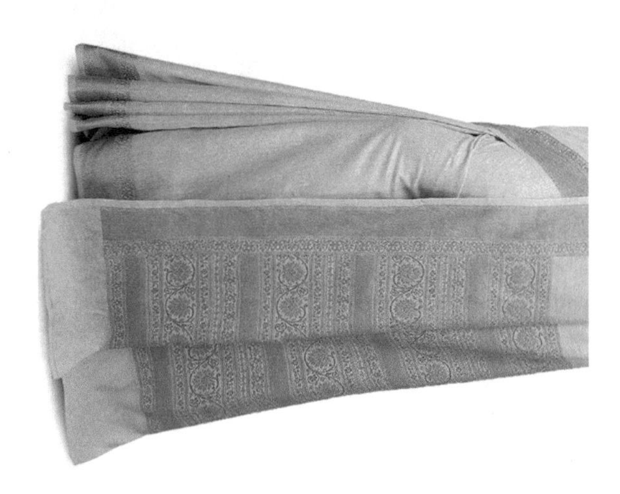

Fig. 24 Muga silk saree

check has four microchecks creating an illusion which is the special characteristics of Kota Coria saree. The design does not appear very prominent on the saree fabric. It is lightweight, open-textured soft feel fabric (Fig. 25).

Uppada Jamdani silk saree

These sarees are popularly produced at Uppada in East Godavari District of Andhra Pradesh in which the designs are perfectly woven along the plain ground fabric without any float on both sides of the fabric. The extra weft is interlaced with warp threads to form the design from left to right and vice versa in such a way that it cannot be pulled off. The specialty of this saree is that the design consists of floral patterns, parrots, and animal designs in bold form are woven by silk/zari (Fig. 26).

Fig. 25 Kota Doria saree

Fig. 26 Uppada Jamdani silk saree

Tanchoi Saree/fabric

This is the product of Banaras in Uttar Pradesh. The fabric fees is very smooth due to the use of satin weave design as basic design. The warp yarn density is very high compared to weft yarn. No floats are visible in back side of the fabric, rather flat in appearance. The back side of the saree fabric appears as a shadow of design of the face side (Fig. 27).

Begampuri cotton saree

Begampuri cotton saree is the traditional saree of Begampur in Hooghly District of West Bengal which can be distinguished by the presence of designs and 'chiur' (designs made by wooden pattyas) technique of weaving in some varieties. Contrast colours are usually arranged in body and borders and some sarees are woven with combination of Dhanikhali origin (Fig. 28).

Cotton Pattu Shawl/Chaddar

Produced in Jodhpur, Bikaner and Jaisalmer Districts of Rajasthan. This specific design is achieved by extra weft technique without use of jacquard or dobby. This

Fig. 27 Tanchoi saree

Fig. 28 Begampuri cotton saree

design is always in contrast colour. Swawls are heavier in nature due to use of coarser yarn. Only geometrical patterns like line, zig-zag, diamond, triangle, etc. are found in the motif/design area (Fig. 29).

Salem Silk Dhoti

This is also a traditional dhoti woven at Salem, Tamil Nadu using silk filament yarns. Speciality of the dhoti fabric is very lustre in nature which is created by flattening the surface with the help of small brass/steel plate on loom itself during weaving. The zari in the border and pallu is also unique. Design appears identical on both

Fig. 29 Cotton Pattu shawl

sides of the fabric. Solid contrast borders on both sides of the dhoti are woven with three-shuttle technique. Many of the times, two different coloured borders are woven like red-green, blue-red, green-orange, etc. (Fig. 30).

Tangail Cotton saree

It is a traditional saree woven in Nadia and Burdwaman Districts of West Bengal. This saree can be identified from extra warp designs in borders and body essentially jacquard is incorporated to achieve the design. Coloured yarns are used in designing the whole body and borders of the saree. Bulging effect in the design areas is found in the fabric due to use of extra yarns in the fabric (Fig. 31).

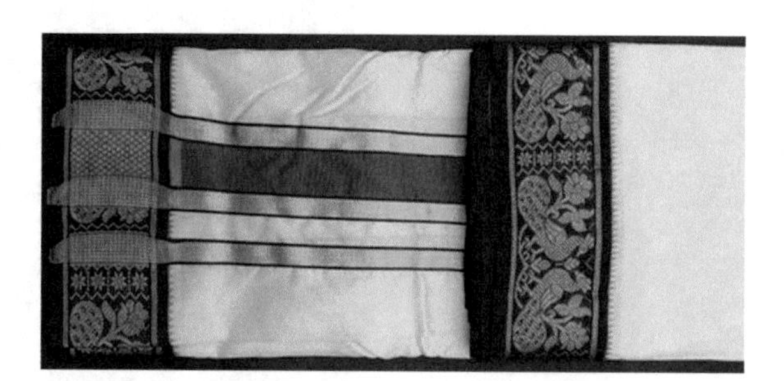

Fig. 30 Selem silk dhoti

Fig. 31 Tangail cotton saree

Venkatagiri cotton saree

Product produced at Venkatagiri in Nellore District of Andhra Pradesh, Venkatagiri saree is not having very compact texture as compared to Balarampuram saree. The borders of the saree have plain zari stripe without extra warp design. Small extra weft design in body and pallu with zari adds beauty to the Venkatagiri saree. Normally, these sarees are light coloured, light in weight and soft in touch since very less designing is done as well as less starch is applied on yarns (Fig. 32).

Tweed

This traditional product produced in various villages of Himachal Pradesh and Jammu and Kashmir in the weight ranges medium to heavy weight fabric. It has a rough surface texture and is produced in great variety of colour and weave effects. The weave design tweeds include herringbones, diamonds, chevrons, cross twills and checks. Most tweeds are colour woven from dyed woollen yarn and some are piece-dyed (Fig. 33).

Fig. 32 Venkatagiri cotton saree

Fig. 33 Tweed

Fig. 34 Shantipuri cotton saree

Shantipuri Cotton Saree

It is also a traditional saree of Santipur, Nadia and some parts of Burdwan Districts of West Bengal. These sarees can easily be identified from the extra warp designs. No other saree will have unique designs like this. Generally, no extra weft buti designing is done in the saree. These sarees are produced with fine cotton yarn with high ends and picks with comparatively stiff starch finish applied on it (Fig. 34).

From the above discussions, it is clear that India has a great diversity in handloom products representing different parts of country. Most of the products use cotton, silk and wool yarns produced in India. The processing technique follows mostly traditional and environment-friendly. Even the processing materials like starch for sizing, natural dye for dyeing, etc. chemicals and materials used are locally available and mostly safe to use. In many of the cases, the dyes used to colour the yarn/fabric are of natural origin available locally. Handloom weavers learnt the operations, processing and skill from their ancestors. Though the products are very unique and special, the production is comparatively slow, and as a result the cost per unit is higher than the power-loom products. However, the handloom weaving community is mostly facing stiff competition with synthetic counterpart and increase in cost of production day by day becomes the challenging phase of the weavers for their livelihood. Also the trend and policies also affect the market of the handloom products. Some advantages nowadays are that the weavers are gaining for better marketing their products using modern online platforms. However, in most of the cases, the weavers of India are very poor and require to lend money from Mahajan/middleman, who got the most benefit by lending money on high interest, purchasing the product at comparatively much cheaper rate and selling them at a higher cost. In some parts of the country, the middleman provides the raw materials, i.e. yarn to weavers. Weavers convert it into fabric/finished saree/product and hand it over as finished goods, and on return a minimum cost of production per unit basis is being provided. Due to lower pay and lower income, many of the weavers are migrating to town/cities from weaving generation old weaving profession. It is indeed a great concern that in near future the tradition may be lost because of such migration from the profession as the full package-of-practice is learned from the ancestors. The weavers not only weave the traditional fabrics but are also in a position to repair and maintain the looms and accessories at their own level. Government policies and incentives will sustain the culture and tradition in future [21, 22]. Although government has year marked yarn production by mill owners in the form of hank for the ease of operation. Government of India offers different schemes to handloom workers and family members like Handloom Weavers Comprehensive Welfare Scheme (HWCWS) which is providing life, accidental and disability insurance coverage to handloom weavers/workers under the components, Pradhan Mantri Jivan Jyoti Bima Yojana (PMJJBY), Pradhan Mantri Suraksha Bima Yojana (PMSBY), Converged Mahatma Gandhi Bunkar Bima Yojana(MGBBY), etc. [1].

As far as the technological intervention on handloom and its products on jute-based products are concerned, some work has been reported in the literature, wherein modifying the existing cotton handloom to suit for weaving coarse jute yarns (Anonymous 2008a). It has also been suggested that with suitable modification [23-28] in processing jute-based warm/winter fabrics [15] to make jacket and wide range of home textiles and utility fabrics for slipper, bags etc. can be woven in handloom [5, 6–8, 14]. The jute-based fabrics are suitable for warm garments like shawls, jacket, etc. and the thermal insulation of the products is comparable to that of woollen garments [16–19] irrespective of its aesthetic look. These jute-based handloom products are very decent in look and because of aesthetic appearance the demand is increasing day by day.

3 Conclusions and Recommendations

India is diverse in traditional handloom textiles and almost all parts of the country have unique design and product ranges. A wide range of handloom material, starting from dress material to furnishing, are being produced from different natural and metallic yarns. The products are very unique and almost not possible to produce in modern power/shuttle-less looms, hence fetches better market value. The livelihood and sustainability of the traditional handloom products require intervention of governmental policies and subsides on procurement of raw material and handloom accessories.

References

1. Anonymous 2019, https://pib.gov.in/PressReleasePage.aspx?PRID=1578522#:~:text=Handloom%20Weavers%20Comprehensive%20Welfare%20Scheme,Bunkar%20Bima%20Yoja

na(MGBBY), Dated 23.02.2021
2. Ahmed K.A, 2015, 'How well do you know Indian handlooms and fabrics', August 06, 2015,
3. https://www.bebeautiful.in/fashion/trends/how-well-do-you-know-indian-handlooms-fabrics, Dated 10.01.2021
4. Anonymous 2008a, Annual Report 2007–2008, National Institute of Natural Fibre Engineering & Technology (Erstwhile National Institute of Research on Jute & Allied Fibre Technology), Kolkata, India, pp.32.
5. Anonymous (2008b) Annual Report 2007–2008, National Institute of Natural Fibre Engineering & Technology (Erstwhile National Institute of Research on Jute & Allied Fibre Technology), Kolkata, India, pp 33–34
6. Anonymous (2011) Annual Report 2010–2011, National Institute of Natural Fibre Engineering & Technology (Erstwhile National Institute of Research on Jute & Allied Fibre Technology), Kolkata, India, pp 15–18
7. Anonymous (2013) Annual Report 2012–2013, National Institute of Natural Fibre Engineering & Technology (Erstwhile National Institute of Research on Jute & Allied Fibre Technology), Kolkata, India, pp 19–20
8. Anonymous (2014) Annual Report 2013–2014, National Institute of Natural Fibre Engineering & Technology (Erstwhile National Institute of Research on Jute & Allied Fibre Technology), Kolkata, India, pp 22–23
9. Anonymous (2017) 'Manual for Indian Handloom Brand Products', Ministry of Textiles, Government of India, pp 1–24
10. Anonymous (2021a) https://www.fibre2fashion.com/industry-article/3759/facts-about-weaving-loom-types, Dated 10.01.2021
11. Anonymous (2021b) http://www.handicraftsindia.org/handlooms/history-handlooms-india, Dated 18.01.2021.
12. Anonymous (2021c) https://www.dacottonhandlooms.in/know-handloom/, Dated 10.01.2021
13. Ahmed KA (2015) How well do you know Indian handlooms and fabrics
14. Sanjoy Debnath AN, Basu RG, Chattopadhyay SN (2009) 'NIRJAFT's Technologies for Rural Development', pp 136–142, Section-V, entitled 'Natural Fibres and Geotextile Applications', Book title 'New Technologies for Rural Development having Potential of Commercialisation', Editor, Dr. Jai Prakash Shukla, Published by 'Allied Publishers Pvt. Ltd.', India, 2009. ISBN: 978–81–8424–442–7.
15. Debnath S (2013a) 'Jute-Based Warm Fabrics', book title 'Diversification of Jute & Allied Fibres: Some Recent Developments', Edited by, Dr. K K Satapathy & Dr. P.K. Ganguly, Published by NIRJAFT, Kolkata, India, 2013, pp 87–98
16. Debnath S (2013b) "Chapter 20 - Designing of Jute–Based Thermal Insulating Materials and Their Properties", pp 499–218, "Textiles: History, Properties and Performance and Applications", Editor: Md. Ibrahim H. Mondal, Nova Science Publishers, 2013. ISBN: 978–1–63117–274–8
17. Debnath S (2014) "Development of Warm Cloths using Jute Fibre", Book title, "Jute and Allied Fibres – Processing and Value Addition: Ed. Debasis Nag and Deb Prasad Ray, (2014), pp. 147–151. (New Delhi Publishers). ISBN: 978–93–81274–41–5 (Print).
18. Debnath S (2015) Chapter 5: Design and Development of Jute-Based Apparels, Book title: 'Handbook of Sustainable Apparel Production', Ed. Subramanian Senthilkannan Muthu, CRC Press, Taylor & Francis Group, USA, April 2015, ISBN 978–1–4822–9937–3, pp. 97–111.
19. Debnath S (2016) Chapter 3: 'Thermal Insulation Material Based on "Jute"', Book title 'Insulation Materials in Context of Sustainability', In: Almusaed A (ed) Publisher: InTech, pp. 45–56. ISBN: 978–953–51–2625–6. (Published online on 31st August, 2016). 10.5772.63223.
20. Debnath S (2017) Chapter 7: Sustainability in Jute-based Industries. Book Title, 'Sustainability in Textile Industry', Part of the series 'Textile Science and Clothing Technology', In: Muthu SS (ed) Springer Nature Singapore Pte Ltd. Singapore, 139–147. Print ISBN 978–981–10–2638–6, Online ISBN 978–981–10–2639–3, DOI https://doi.org/10.1007/978-981-10-2639-3_7. (Published online on 15th October, 2016).

21. Debnath S (2017a) Chapter 3: Sustainable production of bast fibres. Book Title, 'Sustainable Fibres and Textiles', A volume in The Textile Institute Book Series, Pages 69–85, In: Muthu S S (ed). Elsevier Ltd. Duxford, United Kingdom, 69–85. Print ISBN 978–0–08–102041–8, Online ISBN 978–0–08–102042–5, DOI https://doi.org/10.1016/B978-0-08-102041-8.00003-2.
22. Debnath S (2017b) Chapter 15. Advances in Research and the Application of Ligno-Cellulosic Fibres Emphasising Sustainability. Book Title, Textiles: 'Advances in Research and Applications', Pages 319–335, December 2017. In: Mahltig B (ed) Nova Science Publishers, New York, USA. ISBN: 978–1–53612–855–0.
23. Debnath S, Sengupta S, Singh US (2008) A method for producing blended yarn from jute and hollow polyester and method of preparing union fabric and shawl from the said blended yarn, Indian Patent No. 290640, Granted December 14, 2017, Applied July 09, 2008.
24. Debnath S, Bhattacharya GK, Singh US (2009) A blanket from jute-hollow polyester blended bulk yarn and method of preparing the same, Indian Patent No. 310348, Granted March 29, 2019, Filed on August 28, 2009.
25. Sengupta S, Debnath S (2016) 'Jute Based Textiles in Modified Cotton Handloom', annual technical volume of textile engineering division board, vol 1, pp 38–41
26. Sengupta S, Debnath S, Bhattacharyya GK (2008) "Development of handloom for jute based diversified fabrics modifying traditional cotton handloom". Indian J Trad Knowl 7(1):204–207.
27. Sengupta S, Debnath S (2010) "A new approach for jute industry to produce fancy blended yarn for upholstery". J Sci Ind Res 69(12):961–965
28. Sengupta S, Debnath S (2012) "Studies on Jute-based ternary blended yarn". Indian J Fibre Text Res 37(3):217–223

Teaching About "Fibre": Between Art and Contemporary Design

Dorina Horătău

Abstract It is well known that the art of weaving is, originally, a craftsmanship of the handloom. Only very late that "skill" was associated with a design first and then with decorative art, and more recently, in late modernism and postmodernism, with art without distinction of type. What makes this art interesting is its ability to house and communicate symbols—the more interesting, the more they are formally or expressively unique. Nowadays, a special emphasis is placed, in the world, on preserving the authenticity of the physical and cultural place. The teaching method the author proposes is based on a linear and systematic development of successive work stages, which result in a deep and responsible understanding of a subject of study, aligned with current trends in the field. As an artist and pedagogue, the author believes that gradual learning allows students to observe the way in which the "source of documentation" is valued. (Here, the "source" is Mogosoaia Palace, built by Brancoveanu, 1698–1702, an emblematic work of art for the Romanian aesthetic spirit of the seventeenth century.) The documentary material, after a long observation and interrogation of its sources and resources, knows a stage of stylization—meaning a detachment from its "corporality" of what each student considers to be its "inner sense", then transformed into an artistic proposal and discussed at length so that it does not contradict the natural and symbolic data of the source object on one hand, and of the future "receiving matter" on the other hand. The effective work, in the next stage, represents a permanent adaptation to the qualities, as well as to the defects of the material, corrected by the improvisation and the technical solution. The finish, which finally enshrines the newly proposed morphology, shows, in fact, the degree to which each student achieve to understand the extraordinary offer contained in the processed material, to which he gave a new life.

Keywords Teaching fibre art · Heritage works · Teaching methods · Stylistic interpretation · Decorative spirit · Artistic product · Textile design · Artistic experiments

---

D. Horătău (✉)
Department of Textile Arts - Textile Design, Faculty of Decorative Arts and Design, National University of Arts, Bucharest (UNAB), Romania
e-mail: dorina.horatau@unarte.org

M. Á. Gardetti et al. (eds.), *Handloom Sustainability and Culture*, Sustainable Textiles: Production, Processing, Manufacturing & Chemistry,
https://doi.org/10.1007/978-981-16-5665-1_3

Art education, aware of its responsibility in the direction of cultivating and rediscovering values of life that can be very helpful to the requirements of the present, now leans more carefully on its own roots, which can nurture new directions of adaptation of contemporary man to a more intense, fuller level of life. This is why the discipline of weaving, which has recently become an art, deserves the respect of dedication beyond the labour of the hands: in our lives, the thread is a key presence, and its place among our intellectual concerns should be appropriate. Beyond workshops, manufactories or fabric factories, the artist sees the thread as a spiritual connection with others: this is the "destiny" of the art of weaving that teachers try to pass on to their younger colleagues.

It is generally accepted that the art of weaving is, initially, a refinement of manual weaving: from the need to create clothes more and more suited to one's own measure to one's own natural and cultural environment, as well as from the need to provide warmth, comfort and prestige to his own shelter, man narrowed down and "decanted" a sum of manual practices around the textile thread, thus creating a skill from this occupation, then a trade and, finally, an art. Despite the relatively rapid development of the quality of this mastery and its spectacular results (on the one hand special and rare fabrics and garments were made, while on the other hand, exceptional tapestries and embroidery), for a long time, the achievements of textile art were associated with "raw" craftsmanship, that is handicrafts.

Only very late, towards the modern era, approximately from the middle of the nineteenth century, this "skill" began to gain recognition, a "rise in rank": it was first associated with design, then it was part of the broad spectrum of decorative arts. Only much more recently, in late modernism and especially in postmodernism, it has been received and valued as an art without distinction of the aesthetic genre. Let's say that today, all artists take on the status of "visual artists", a generalization that we think, however, to be to some extent unfair to the authors of objects crafted manually, because the touch or weight of a material, if well mastered, adds value to a work, and this contribution is too little observable when the lover of art is reduced to the status of a simple image consumer. The touch (be it only anticipated by the spectator-visitor of an exhibition) can only be imagined, as well as the smell associated with textile fibre, not to mention the total holistic effect that a large tapestry, for example, has on the one who encounters it, and is impossible to be conveyed through an image; finally, the image does not do justice to the object created from thread because it too seldom shows the way in which such a work of art communicates with an architectural environment, with a place that, frankly, enriches it. What makes the art of the thread interesting is precisely its ability to embed and communicate symbols through this implicit and explicit materiality, all the more interesting as the morphologies under which it is revealed to us are expressively unique.

Unlike the way in which the learning process, the pedagogue–student relationship unfolded no more than 20–30 years ago, the changes that have taken place today are dramatic: the young contender for the status of the artist is more and more informed, more demanding and more willing to be seduced by the pedagogical offer; unfortunately, often lacking preparation and with an overflowing taste for any type of fast food, including the intellectual one. In the field of arts as well, before the professor, he/she looks for the role model, the older colleague, but not so old that he/she cannot question their options, the proposed topics and themes, their achievements and their way of teaching. This request is not entirely illegitimate, because only research-type work, always in the sense of searching for and discovering new things, allows both parties to maintain their desire for labour, their taste for work and their respect for art. Fortunately, those who manage to overcome the first temptations are, later, willing to go in for the long haul and to undergo the difficult evolution through art, which begins, for the textile students of the University, with tiring but necessary hours of documentation and applied understanding of what they see, continues in the specialized laboratories of the university, which offers them human and material support that they will not have later, and ends in the annual year and group exhibition and, later, in the first exhibitions in the public space, when their entire education will be judged and appreciated. But it must also be said that the Romanian art of weaving also offers satisfaction: not only because it comes from a prestigious past, proving a high culture, worth the search effort, but because the works of many Romanian weavers enrich, enliven, in the country and abroad, hundreds of residential spaces, and the names of many of them are part of the cultural heritage of Europe: Ritzi Jacobi, Ana Lupas, Theodora Stendl and Ariana Nicodim. Each new day of teaching should include the urge, barely uttered, to the students to follow in their footsteps.

But not just this: the need to know and develop a healthy and learning environment for all people must be a major goal for any mature society, aware that in its bosom there are simple workers, intellectuals, artisans, but also artists: everyone can participate in the generation of a good quality social and cultural environment, the condition being that each brings their own "best contribution" for this cause.

Nowadays, under the pressure of globalist culture and arts, a special counter-emphasis is placed in the world on the presence of local artistic styles, the stimulation of community popular cultures and the specifics of craft industries. All these "instances" having the advantage, on the one hand, of preserving the authenticity of the physical and cultural place from which they are derived and which they represent, on the other hand, the mentioned fields bid on their ability to reach the idea of the present by assuming the concept of contained sustainability, one of the major objectives of today's society.

A choice that has returned frequently in recent years is the one that visits local artistic styles, especially those that allow accessible documentation and that potentially contain seeds of "development" towards an artistic object with multiple valences. Easy to consult even physically, given its relative physical proximity to Bucharest, therefore offering immediate accessibility, the Mogosoaia Palace (built by Brancoveanu, 1698–1702) is, at the same time, an offer full of resources, which can provide a wide range of options for those interested in a fertile dialogue with decorative art. Moreover, thanks to a high-quality restoration, carried out under the patronage of Princess Martha Bibescu, the decorative details were well recovered and their role in the economy of the entire ensemble was reaffirmed. Therefore, the "source" of information offered to students benefits, besides the prestige of an architectural example of aristocratic taste, emblematic for the Romanian aesthetic spirit of the seventeenth century, from the fruit of an exceptional restoration, which guarantees the expressive, material and symbolic authenticity of the contained forms.

The coordinating professor considered that there are at least three important reasons for such a topic to be approached.

The first reason is, for the history of Romanian art, the so-called Brancovenesc era is the first to connect to European modernity of the time, in the sense that a series of non-religious elements were introduced in art, and the idea of beauty, in the sense of pure aesthetic pleasure, detached from the religious meaning, is accepted in court life, art intertwines with life (see architectural transformations, a series of forms that emanate concreteness and sensuality, the taste for delight appears as a way of cultivation and education). This last quality, of transcending the boundaries between the living and the aesthetic, was remarked, moreover, by a Romanian researcher of the Baroque on the Romanian territory, speaking of the Mogosoaia Palace, in particular, of the Brancovenesc spirit in particular: "If in the Renaissance, a certain religious or mythological appearance had always been present/preserved in works of art, the baroque was established (…) by a glorification of the mundane (…), by the praise brought to the human body as biological, as exposed naturalness, worthy of admiration. (…) If, to the above-mentioned observations we will add the Renaissance geometry in the disposition of the buildings, but totally baroque in spirit, from the point of view of the interlacing between something made and a living thing, between vegetal and artificial, between Apollonian and Dionysian, we will have to accept that we are facing one of the most relevant proofs of the new spirit that had penetrated the culture of these places, at least in the spirit of the cultured people." (2)* ([2]: 69) We do not even think that we need to insist on this spirit of prosthesis or synthesis which is omnipresent in contemporary art, and which continues, intellectually, but also in terms of material-sensitivity, this play between nature and culture, between living and artificial, between gross existence and accused invention which, historically speaking, we can find in any manifestation of the mature baroque. The second reason is that the era was recovered by local history as one of "definition" and connoted with a sense of dignity, national and international affirmation of local pride. Its promoter, Constantin Brancoveanu, was suddenly a cultured and very rich character, and he has made several architectural works of great local significance—Mogosoaia Palace, influenced by both East and West, as the same scholar tells us, being a representative in this sense.

Finally, the third reason is that, thanks to its freely asserted luxury, the era was reconsidered, fragmentary and selective, not only by the history of twentieth-century Romanian architecture but also by a series of contemporary designers and artists who were inspired by the artefacts of the time to recreate objects of "vintage art".

Very few of us realize that every artistic or design product depends, and ultimately is the result of a whole series of refinements, the ultimate meaning of which would be the revelation, awareness of the artistic act, its autonomy and its ability to communicate to the viewer the idea of sensitivity and beauty of the external world surrounding us, but which is less and less felt, more and more vaguely perceived by contemporary man. But the recovery of the real desire to create, in the case of learning students, as well as the development of a sense of aesthetic sensibility in the art lover/spectator, increasingly threatened by the "thermal death of the senses" (1)* Konrad Lorentz) are, finally, the main objectives of the tutor-teacher.

The object "backed" by the documentary material, after a long and careful observation and interrogation of its sources and resources, both visual and textual, undergoes the first stage of stylization, i.e. an extraction from its initial, compact "corporality", revealing and setting apart an aspect of it, and a much more precise formulation of the area of interest, accompanied by what each student considers essential. The fragment of a decoration sculpted from the palace loggia, the princely eagle, for example, is reconsidered by virtue of its quality of being able to be stylized in a decorative-contemporary spirit, to be re-synthesized in a formula that addresses an entire habitat—from the mosaic decoration of a floor to a type of textile decoration attached to some furniture objects, designed in the same spirit. This internal quality discovered and tuned with the sensitivity of the "researcher" is then transformed into a draft artistic proposal, discussed at length, so as not to contradict the natural and symbolic data of the source object, on the one hand, nor those of the future "receiving matter", on the other hand, in order to make an agreement in principle between the idea material 1 and the finished object 1, in order to be able to anticipate, to a good extent, the problems that will later require solving. Effective work, in the next stage, is a permanent adaptation to the qualities as well to the defects of the material, corrected through an improvisation-adaptation process with the help of the most suitable technical solution. The completion, which ultimately establishes the newly proposed morphology, shows, in fact, the degree to which each student understands the extraordinary offer contained in the material process with the help of which he/she gave a new life to a new reality.

The proposed teaching method is based on a linear and systematic development of successive work, in stages, so that through a sensitive, conscious and committed work of the student it can result in a deep and responsible understanding of a subject of study, aligned with current trends in the field. As an artist, pedagogue and careful researcher of textile art, I believe that gradual learning allows students to observe the path in which the "source of documentation" chosen is gradually understood in its deepest sense. And I usually appreciate the students' option to be inspired, in many situations, by the repertoire of old and very old arts.

Of course, there is a need, also from the students, for a conscious and well-justified relationship to the source of inspiration because only by discovering the real / truthful essences of a previous manifestation can it be continued correctly and nuanced, developed without the danger of misunderstanding and misinterpreting it or transposing it into a possible kitsch element.

1 Overview of the Topic

A topic offered to students must be given with great deference with a process adapted to learning, often even personalized. The staging of this process doubles, in a way, the one presupposed by the labour of the form that will become, at the end of the search, a textile object with artistic quality. Therefore, the guiding pedagogue segments this learning relationship in a series of chronologically successive moments, specific to the pedagogical artistic act transposed in practice, a discourse seen as a way of refining a vision, in which each segment actually describes a step, a more intimate understanding of the aesthetic act, seen as a pathway from the inspirational raw body to the finished artistic object.

For the above reason and also because the textile art is an art of "sincerity" (neither the thread nor the work on it can be falsified, both require direct and committed involvement from the "practitioner"), an "open discourse" is proposed, a visual anatomy of the artistic act, in which one can observe the "course", the becoming of the work, from the source of inspiration to the practical finality of the artistic object or of the textile design object.

Presentation of the study stages for the topic textile design "Brancovenesti Palaces—Source of Documentation and Creation in the Field of Decorative Arts and Textile Design"

Working steps	Study discipline	Description of work stages/Work technique
Stage 1	Two and three dimensional representation techniques	A. Making hand-drawn drawings in the in-situ documentation campaign at the Culture Center "Brâncovenești Palaces at the Gates of Bucharest", Mogosoaia Work techniques: paper work sketches made with pencil, charcoal, count, colored pastel, mixed techniques, as well as various combinations of the student's choice B. Assembling the documentation in theme plans Working techniques: textile collage (paper, various textiles)

(continued)

(continued)

Presentation of the study stages for the topic textile design "Brancovenesti Palaces—Source of Documentation and Creation in the Field of Decorative Arts and Textile Design"		
Stage 2	Stylistic Study Composition for Tapestry Weaving / Printing	Compiling a collection of textile design for clothing or interior decoration A. Selection and composition of representative theme pages (Fig. 3) B. Selection / composition of some decorative elements taken from the theme pages and creation of a collection of textile design (projects—open field compositions or in association with tapestry project), which follows the relationship between a major project and one or two coordination projects at the major project. Working technique: digital environment and projects made on paper and water colors (Fig. 4 and 5) C. Composition of the textile design collection for presentation in the folder (theme / subject approached, atmosphere page of the textile design collection, series of major projects in relation to projects-coordination 1, 2 virtual montages
Stage 3	Transpositions into materials specific to tapestry weaving/printing	Specialization tapestry weaving A. Compositions for tapestry weaving Working techniques: haute lisse Purpose: tapestry for interior decoration

This work describes a part of the teaching approach used within the Department of Textile Arts—Textile Design, Faculty of Decorative Arts and Design of the National University of Arts, Bucharest. According to the general presentation on the university website, *B.A. studies offer general knowledge and skills in the field of textile art and design: textile arts as artistic expression (fibre art, object, installation) in textile medium; textile design for interior decoration collections; textile design for clothing collections (fabrics or prints). M.A. studies in ambient textile arts offer students the opportunity to thoroughly pursue a proposed project, with an emphasis on: creativity, original personal contribution, investigation, interpretation and innovative experiment. The activity of professors and students takes place in study/design spaces/workshops, technological spaces—laboratories for transposing into textiles the creations, experiments and applied research topics in educational activity (textile printing, tapestry-weaving and computer image processing* (http://unarteorg/departamente/textile/).

If in the first year the study focuses on an introduction to the main specialized issues of the field, in the second year the necessary steps are taken towards understanding and deepening personalized working methods. The stylistic interpretation of some symbols or decorative motifs from different historical periods is one of the reference study directions from the curriculum, as a specialized discipline called *Stylistic Study Composition for Tapestry Weaving/Printing*. Students from the two specializations, tapestry-weaving and textile printing, go through a theoretical documentation, as well as a practical one, with drawings, free colour sketches after albums and books from libraries or after heritage works found or exhibited in museums. It is an exercise in which they learn how to extract, how to select a decor, how to choose a graphic element, and a decorative motif so that later they have the ability to make and complete a collection or a series of unique projects for a certain type of interior space, piece of furniture or for clothing so that it is up-to-date and keeps up with the proposals of decoration and contemporary textile design. At the same time, the idea of composition involves achieving intermediate stages of the final work which can be constituted, by itself, in objects (drawings, sketches, paintings) with independent plastic value.

The role of the guiding professor is very important from this stage: most often, in front of a generous source of inspiration, the student has an exalted, interested attitude, sometimes almost dogmatic towards what he/she would like, already, to complete, to "conclude" in a work. He has almost a vision of the final object. Fortunately for him, the professor guides him, through dialogue and discussion of the limits given by the concrete qualities of the observed "sources", to a first definition of the area of interest, then to a second, which either details or offers complementarity to the first so that a double perspective, at least, would urge the student to a critical exercise towards his/her own possibilities of expression. The linear continuity of an element of Brancovenesc style decoration—the shape of a bird from an openwork decoration—presents, for example, some variations of thickness or suggestions of volumetric contrast. The student can opt for an integrative project, which is based, plastically, on this formal element. The professor, however, will draw the student's attention to the fact that when this formal lightness, expressed by a stylized drawing will decorate the tile floor, it will have to be imagined taking into account the volume and lighting of the room; while if the same shape will be printed on an upholstery, it will need a series of processing and adaptations. Consequently, the design for the floor can be appropriately transposed into the mosaic or charcoal technique. When, however, the same linear-source decoration is addressed to a lampshade, perhaps an interpretation like watercolour might seem more appropriate. The difference, as well as the observation of the complementarity of these projections, can provide the student with adequate guidance regarding the way in which the source of an image can be continued, forced or denied, depending on his/her own visions.

2 Presentation of Work Stages

In the second year of the Bachelor's degree, the following disciplines provide practical activity in study workshops and laboratories:

- Two- and three-dimensional representation techniques;
- Stylistic study composition for tapestry weaving/printing;
- Transpositions into materials specific to tapestry weaving/printing.

In the first semester (academic year 2019–2020) of the second year of B.A. studies, for a better understanding and deepening of the specialized field, the students were offered the topic *Brancovenesti Palaces—Source of Documentation and Creation in the Field of Decorative Arts and Textile Design.*

Under the umbrella of this major theme for the semester, the coordination of the activity took place in three successive stages, during which the objectives of the disciplines listed above were pursued.

In the first stage, related to the discipline of *Two- and Three-Dimensional Representation Techniques*, an in situ documentation campaign was carried out at the Culture Centre "Brâncoveanu Palaces at the Gates of Bucharest", Mogosoaia. The subject of the work was to identify specific motifs or compositional schemes, which the students will interpret in a documentation notebook and will make works inspired by the context. Emphasis was placed on the elements of authenticity, major in this architectural and historical ensemble: the Renaissance-type planimetry of the ensemble, which could motivate certain rationality of the plastic discourse; the insertion (of baroque type) of the construction in nature, which could be an argument for the poly-sensory quality of a proposal, given the concreteness very close to the formal physical aspect of some sculptural details—palmettes, addorsed birds, vegetal decorations; the luxurious aspect of the residence, which could be kept as the dominant note of creation, either perceptibly by the magnitude of a proposal, or by the special care for the realization of some details; a certain pictoriality of the ensemble, which can harmonize very well with the aggressive visuality of contemporary plastic art, dominated by "the thing that wants to be seen". Two aspects were pursued in essence: first of all, the omni-presence of the vegetal element, which, practically, envelops, covers the architectural construction in a lace ribbon of great finesse, aspect to which the effective placement of the building in the middle of nature majorly contributes, and secondly, the collaboration of different types of materials and tectonic textures present in this architectural mini-work: simple stone, marble, brick, in the most varied forms—cut, cast, sculpted, round, square or rectangular, rough or polished, coloured or grey. These aesthetic dimensions, in turn, according to the committed and motivated selection of each student, will become the basis of artistic or textile design projects for contemporary decor or clothing.

The working techniques used were: paper sketches made with pencil, charcoal, conté, coloured pastel, mixed media, photography; various combinations of "information and grading" were also allowed at the student's choice. It is worth emphasizing the direct communication with the source material, in the sense that the students, immersed in the immediate vicinity of the environment dominated by the palace, had the opportunity to observe and then choose their favourite "motif" first based on a convenient subjective impression, in the idea of being completely attached to the chosen "real segment": the materiality and morphology of a detail could be followed at different times of the day, which allowed "feeling" them under different shadows and lights, a very important aspect in the case of this architecture, and also because, in itself, this feeling says something very precious about the capacity of a historical object to become a source of interest for the contemporary spectator; in this case, an informed and interested one (Figs. 1 and 2).

The second factor worth mentioning refers to the way of "capturing" this image material—an impressive number of "notations" were made ad hoc, through special ways of "writing": pencil on paper, drawing on the board or simple minimal sketches made in coloured pencil on notebook pages, "contributions" made either from lying on the grass, or from the steps of the loggia, or simply in the upright position in front

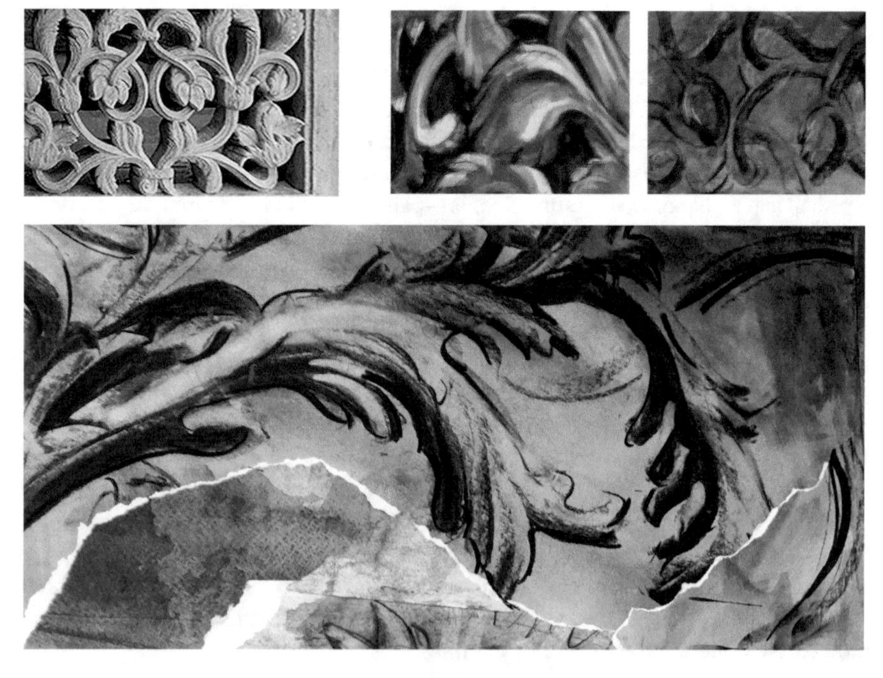

Fig. 1 Photo and sketches/drawings in coal and conte on toned paper; student-author Delia Ciubotaru, second year, licence, coordination: PhD Associate Professor Dorina Horățău, 2019–2020. *Photo credit* personal archive Dorina Horățău

Fig. 2 Photo and sketches/drawings in coal, conte and watercolour on toned paper; student-author Claudia Davidescu, second year, licence, coordination: PhD Associate Professor Dorina Horătău, 2019–2020. *Photo credit* personal archive Dorina Horătău

of an ornament, each suggesting, in fact, the essential way of the youngster's relating to his area of interest in the Palace.

The second stage, related to the discipline *Stylistic Study Composition for Tapestry Weaving/Printing*, aimed at compiling a collection of textile designs for clothing or interior decoration.

This stage, which takes place in the workshop, based on the sketches selected from the individual documentary material, mainly aims at the selection and composition of representative theme pages (35 cm × 50 cm format). Based on this selected material, it is proposed to compile a series of "stand-alone compositions" that includes the sketch—hand drawing in relation to other related materials/dialogue partners: surfaces of colour-processed paper or series of textile scraps, to which the sketch relates in terms of colour, texture and materiality. Thanks to this exercise, the chromatic range of the future proposal is intuited, and the two main colours are defined; we have a first starting point for a "collection" of decorative elements that can be later embedded in textile design projects of greater scope. One or more sketches considered by the student as relevant for the way he/she perceived the approach to the chosen motif are arranged together with a variant of material, in the idea that the identified morphology can acquire or can be associated with a concrete sensitivity. The exercise is also a compositional one, with an aesthetic and pedagogical purpose of its own, because the transposition of the two approaches on a single board creates contrasts or similarities often surprising, promising, but it also functions as a

Fig. 3 Theme pages; student-author Claudia Davidescu, second year, licence, coordination: PhD Associate Professor Dorina Horătău, 2019–2020. *Photo credit* personal archive Dorina Horătău

"palette" or "table of contents" because it can be used as a "base", as a starting point for all subsequent work phases (colour palette, textures of woven or printed surfaces, observation of different types of textiles, identification or specification of an accent composition element) (Fig. 3).

From this series of plans that trigger an idea or proposal, a working direction is chosen—3–5 boards/compositions are selected, which are compatible not only in terms of colour range but also the formal "consonance" of the decorative elements, after which, depending on the "demand" of each decorative surface and its inner structure, a "particular imprint" is proposed, given by a certain circulating motif. The problems of stylization, size and arrangement of the decorative element taken over or interpreted are related to the work technique chosen for the transposition into material either within the specialization of tapestry-weaving or the specialization of textile printing. The fragment of a stylized vegetal element is redrawn in another open-field composition. This fragment is repeated/serialized according to a pre-existing compositional scheme, generating a composition that will be used as a project for the fabrics intended for the interior design project—textile wallpaper, fabric for sofa, armchair, etc.—a series of projects that raise the issue of coordination: alternative projects designed as "expression partners" for an interior or an outfit (Fig. 4).

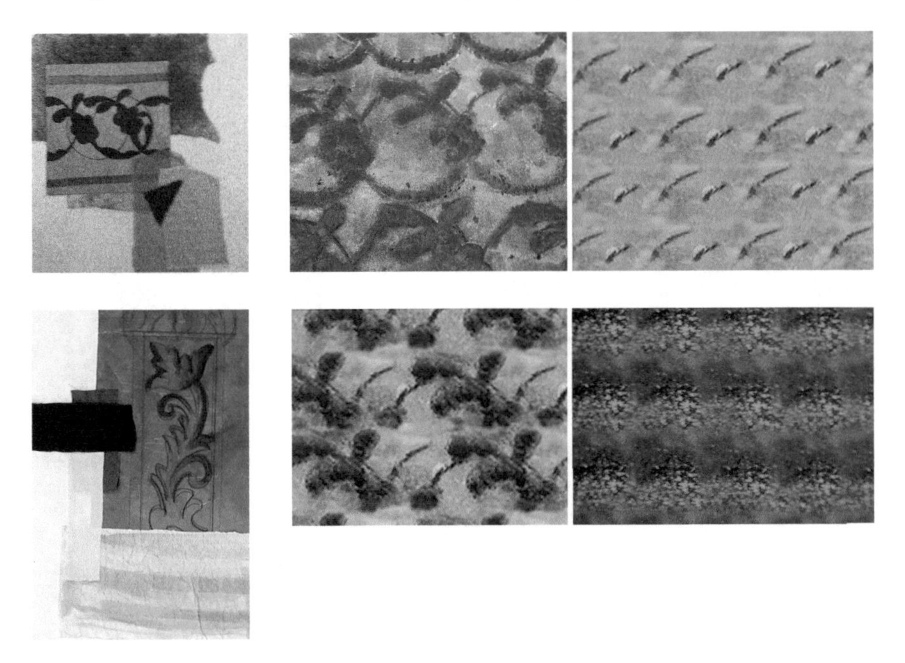

Fig. 4 Theme pages (on the left) and fabric projects (major project and coordination project). Projects made with tempera colours; student-author Claudia Davidescu, second year, licence, coordination: PhD Associate Professor Dorina Horătău, 2019–2020. *Photo credit* personal archive Dorina Horătău

The major project always has one or two proposals for coordination projects "attached", together with which they can form a unitary decorative ensemble. Thus, if the major project supports the main piece of a room (a sofa that occupies a privileged place), the coordination project will refer to the "secondary" pieces of this decorative composition—its "object" will be the curtain, the upholstery of the chairs and textile accessories. The *pièce de résistance* of an interior defined around the thread is, naturally, a tapestry that will assume the status of star object of the room not only because it will connect all the soft components of the space, but because it itself is, in essence, the refinement and the superior individualization of an inner and formal aspect which, as a "background", is alive and dynamizes the entire artistic component of the respective house/room (Fig. 5).

By a particularization of a leitmotif intuited as the spiritual and material presence of that place, by treating it in a very special way, which is related to the uniqueness of the approach and treatment of that motif, by its composition in a woven space, whose rules themselves are unique, the central object reaches the status of art. The tapestry is the central piece of the proposed interior ensemble.

Fig. 5 Fabric collection presentation page for interior decoration and integrated tapestry; student-author Claudia Davidescu, second year, licence, coordination: PhD Associate Professor Dorina Horătău, 2019–2020. *Photo credit* personal archive Dorina Horătău

The series of sketches attached in the independent compositions has the role of triggering the development of a search that leads, finally, to the creation of a collection of interior textile designs with the integrated tapestry. Each collection usually contains 5–7 major projects and 1–2 coordination projects for each major project, individualized colour palette, atmosphere page. The design of the collection is usually made following two directions: one that allows working with watercolours, tempera on the board, in the workshop, the second using digital design. Working with traditional artistic materials offers the possibility to faithfully represent the visual qualities of textile materials; in digital design, many more options can be obtained, variations of the chromatic ranges regarding the disposition of decorative materials and even selection and distribution on different surfaces (Figs. 6 and 7).

The requirements of the discipline *Stylistic Study Composition for Tapestry Weaving/Printing* include:

- putting together a board to present the theme, which, basically, recapitulates the entire artistic approach from the origins until the present. Thus, visualizations of the projects can be operated in different spaces or contexts, which can approximate the presence of the decorative element in that place (Figs. 8 and 9).

Fig. 6 Fabric collection presentation page for interior decoration and upholstery cardboard for tapestry; student-author Delia Ciubotaru, second year, licence, coordination: PhD Associate Professor Dorina Horătău, 201–2020. *Photo credit* personal archive Dorina Horătău

- compiling a catalogue of the textile design collection containing data on the theme/subject approached, the atmosphere page, a series of major projects in relation to the coordination projects, as well as virtual montages on a character (clothing) or in a space (interior decoration), a technical file and a presentation of laboratory samples (textile material and dyes used, technical data about the fibres/type of fabric, etc.).

Working techniques used are mixed media (tempera on paper or collage with paper and textile material) and digital design. Most often, such a catalogue ends up being an artistic object in itself—when possible, it includes all the phases of documentation for an object or a theme, and in it are stored not only the written notes but also the drawings and the sketches related to the different stages, which make it possible to observe the amazing variety of approaches on the topic, as well as the sometimes surprising, other times predictable answers, sometimes conventional but also, in many situations, exciting challenges both to the "source" and to the process that their labour involves (Fig. 10).

Fig. 7 Fabric projects for interior decoration and integrated tapestry; student-author Claudia Davidescu, second year, licence, coordination: PhD Associate Professor Dorina Horătău, 2019–2020. *Photo credit* personal archive Dorina Horătău

The third stage of work involves refining the results obtained in the first and second stages in order to select viable projects for transposition into textile material in the laboratories of the department. This stage corresponds to the discipline *Transposition in Specific Tapestry/Printing Materials.* For the effective transposition in the material (printing), a cliché is made for a project of the open field composition type with a copy of the drawing in order to obtain the screen with which the material is subsequently printed. In the case of the tapestry project, the selected project is made at a 1:1 scale, of the desired size; it is possible to make, for ease of work, an upholstery cardboard in the form of a digital print (there is also the traditional, manual version, which involves a higher consumption of time and materials). Once the cardboard is made, there is a "pause", another surface that contains a simplified drawing of the entire composition, a kind of overlay, on which the colours in the basic project are marked with numbers, so that the colour transposition would maintain the compositional accuracy of the project (each digit corresponds to a certain colour tone) (Figs. 11 and 12).

Fig. 8 Theme presentation page; student-author Claudia Davidescu, second year, licence, coordination: PhD Associate Professor Dorina Horătău, 2019–2020. *Photo credit* personal archive Dorina Horătău

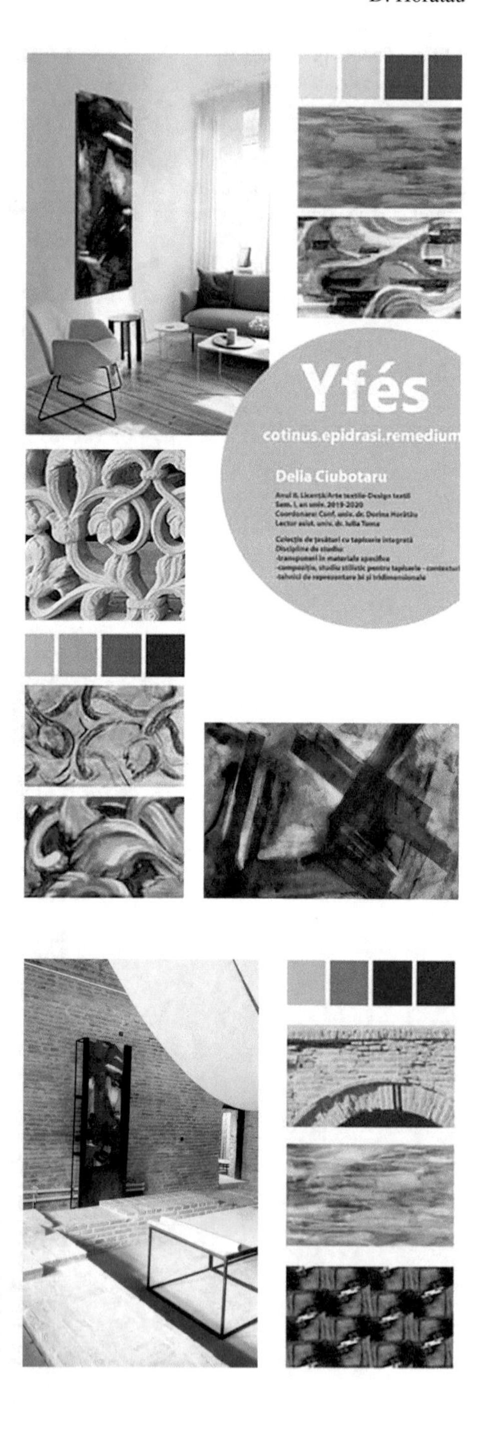

Fig. 9 Theme presentation page; student-author Delia Ciubotaru, second year, licence, coordination: PhD Associate Professor Dorina Horătău, 2019–2020. *Photo credit* personal archive Dorina Horătău

Fig. 10 Catalogue pages of design textile collection; student-author Claudia Davidescu, second year, licence, coordination: PhD Associate Professor Dorina Horătău, 2019–2020. *Photo credit* personal archive Dorina Horătău

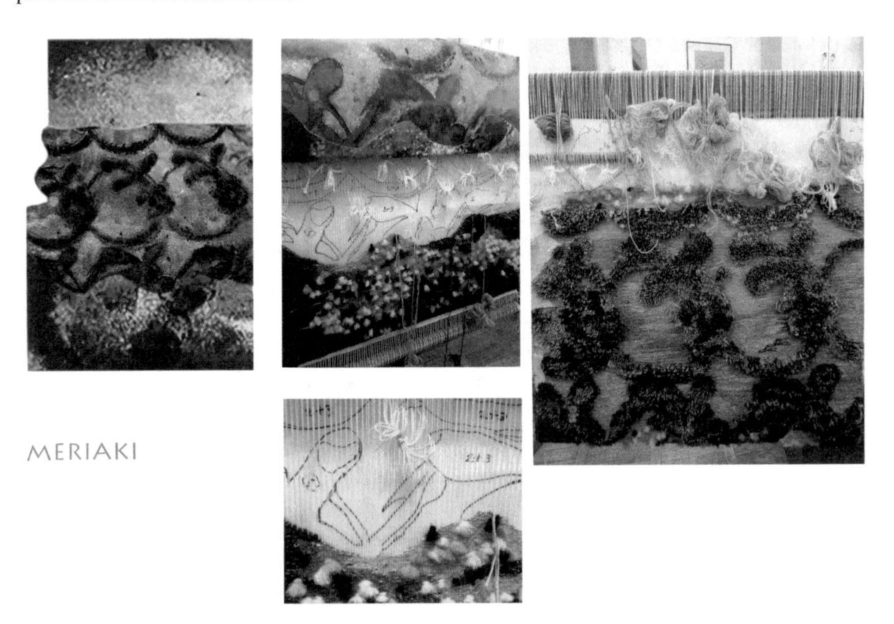

Fig. 11 Upholstery cardboard for tapestry (on the left) and stages during worktime on the haute-lisse loom; student-author Claudia Davidescu, second year, licence, coordination: PhD Associate Professor Dorina Horătău, 2019–2020. *Photo credit* personal archive Dorina Horătău

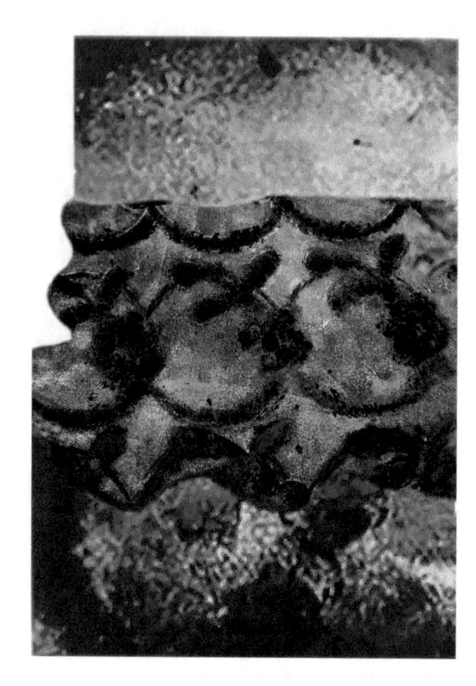

Fig.12 Project/upholstery cardboard for tapestry (on the right) and fabric projects for interior; student-author Claudia Davidescu, second year, licence, coordination: PhD Associate Professor Dorina Horătău, 2019–2020. *Photo credit* personal archive Dorina Horătău

Depending on the required dimensions and the technical impositions of the tapestry, the type of frame is selected, on which the warp is fixed—the vertical threads that support the actual fabric (the weft). If the upholstery contains protuberances, the warp is looser, with thicker cotton thread (3 threads per cm), while for haute-lisse tapestry woven with thin wool thread the warp is tighter (4–5 threads per cm) (Fig. 13).

The warping operation takes place in the laboratories of the Department, and it is done by the technicians together with the student-author. In parallel, depending on the requirements of the project, the colours of wool and the type of fibre are chosen, they are dyed by the technician, with the collaboration of the student, under the guidance of the coordinating professor, then it is proceeded to the actual work, which the student carries out in collaboration with technicians.

....

To exemplify this material, two works were selected: fabric collections with integrated tapestry from the tapestry-weaving specialization.

Work 1

Title: **Yfés**

Student-author: Delia Ciubotaru

Fig. 13 Haute-lisse tapestry during work (modes of warping on loom, fibre types, fabric types); student-author Claudia Davidescu, second year, licence, coordination: PhD Associate Professor Dorina Horătău, 2019–2020. *Photo credit* personal archive Dorina Horătău

Second Year, B.A. studies, specialization: tapestry—weaving

Coordinator: Assoc. Prof. Dorina Horătău, PhD

Working technique: haute-lisse tapestry on vertical frame (haute-lisse fabric, Persian knot, sumak)

Working materials: thick wool, thin wool, vegetable silk, cotton

Dimensions: 80 cm × 185 cm

Year: 2019–2020

An approach that emphasizes the contrast/relationship between two types of textures existing on the body of the historical monument building.

The brick wall has an apparent texture, defined by the repetitive horizontal geometry and the warm chromatics of the burnt clay.

The tracery of the stone railing, on the other hand, has a dynamic dictated by the representation of the S-movement of the stylized vegetal element. Freehand sketches made in charcoal on a vibrating toned background manage to capture a secret dynamic breath of these textured surfaces of the building facade. The final result is an abstract composition that synthesizes / cumulates the two directions, transforming them into a geometric gesture-element (intersected Z-shape) (Fig. 14).

Fig.14 Presentation pages of the used approach for the composition of the upholstery tapestry project; student-author Delia Ciubotaru, second year, licence, coordination: PhD Associate Professor Dorina Horătău, 2019–2020. *Photo credit* personal archive Dorina Horătău

Fig. 15 Haute-lisse tapestry during work (modes of warping on loom, fibre types, fabric types) student-author Delia Ciubotaru, second year, licence, coordination: PhD Associate Professor Dorina Horătău, 2019–2020. *Photo credit* personal archive Dorina Horătău

The composition was made on paper and with watercolours, after which photos were taken and image interventions in photoshop. The final composition selected (tapestry cardboard) was printed in colour to make a tapestry for interior decoration (Fig. 15).

The type of fabric (mixing of haute-lisse fabric with Persian knot with relief between 1 and 1.5 cm and irregular floating of thread over the haute-lisse fabric) and the contrasting chromatic range complete the spontaneous gesturalism dictated by the compositional representation. The initial study subject seems to become a pretext for representing a state or the inner self of the student-author. It is worth mentioning that the tapestry was mostly made by the student-author, with minimal help from laboratory technicians and was completed during approx. 6 weeks of work, a commendable performance. The authentic weaver artist (whom I would like to believe that this student forebodes) means strength, dedication, involvement, endurance, motivation, skills hard to find today (Figs. 16 and 17).

Work 2

Title: **Meriaki**

Student-author: Claudia Davidescu

Second Year, B.A. studies, specialization: tapestry—weaving

Fig. 16 Completed haute-lisse tapestry (fibre types, fabric types); student-author Delia Ciubotaru, second year, licence, coordination: PhD Associate Professor Dorina Horătău, 2019–2020. *Photo credit* personal archive Dorina Horătău

Coordinator: Assoc. Prof. Dorina Horătău, PhD

Working technique: haute-lisse tapestry on vertical frame (haute-lisse fabric, Persian knot, sumak)

Working materials: thick wool, thin wool, twine wool, vegetable silk, cotton

Dimensions: 95 cm × 125 cm

Year: 2019–2020

The theme takes vegetal decoration elements from the door or window frame, as well as from the stone tracery of the building's railings, a historical monument. The hand sketches are made in charcoal or conté on watercolour toned paper. The drawing of the sketches follows a representation of the form, as well as of the volumes dictated by light and shadow. A volute taken from a representation of a decorative element becomes a fragment symbol with partial repetition on a slightly symmetrical surface in a balanced composition that gives the impression of stability (the bottom of the composition occupies three-quarters of the compositional surface). The compositional framing is irregular precisely in order to break the serialized repetition of the decorative element—the volute fragment. The composition was made on a board with tempera and acrylic colours for the faithful representation of the

Fig. 17 Presentation page and ambiental proposal for the completed tapestry; student-author Delia Ciubotaru, second year, licence, coordination: PhD Associate Professor Dorina Horătău, 2019–2020. *Photo credit* personal archive Dorina Horătău

textures, after which fragments with collage were recomposed; subsequently, everything was photographed, the image being processed using the computer (Fig. 19). The final result was colour printed in 1:1 scale to "function" as a tapestry board. It was decided to create a tapestry because it gives the impression of oversized velour and continues the theme of the Brancovenesc style (Fig. 20).

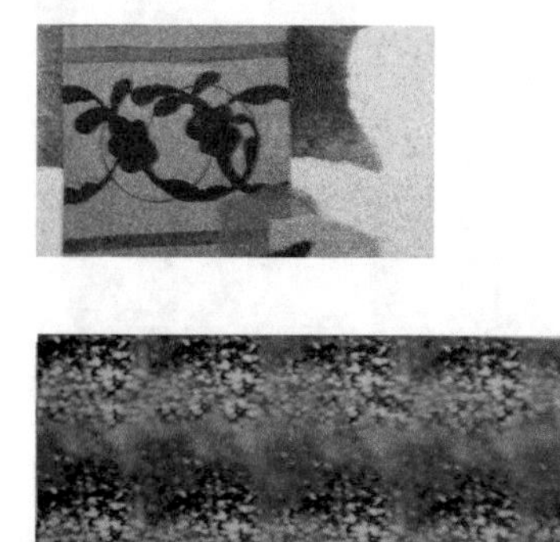

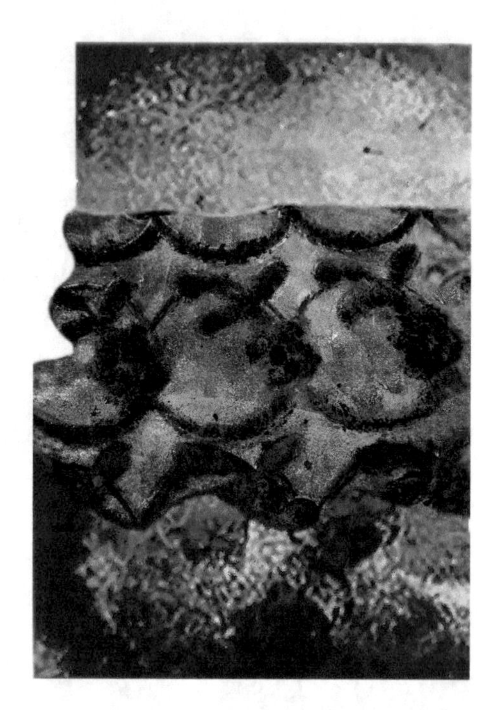

Fig. 18 Decorative element taken over and interpreted and context projects and tapestry; student-author Claudia Davidescu, second year, licence, coordination: PhD Associate Professor Dorina Horătău, 2019–2020. *Photo credit* personal archive Dorina Horătău

Notes

1. Lorenz, Konrad, *Civilized Man's Eight Deadly Sins* (Romanian translation by Vasile V. Poenaru), Humanitas, Bucharest, 2007, chapter "The Death of the Senses", pp. 39–51.
2. Hostiuc, Constantin, *The Romanian Baroque, Gestures of Authority, Answers and Echoes*, NOI Media Print Publishing House, Bucharest, 2008, p. 69.

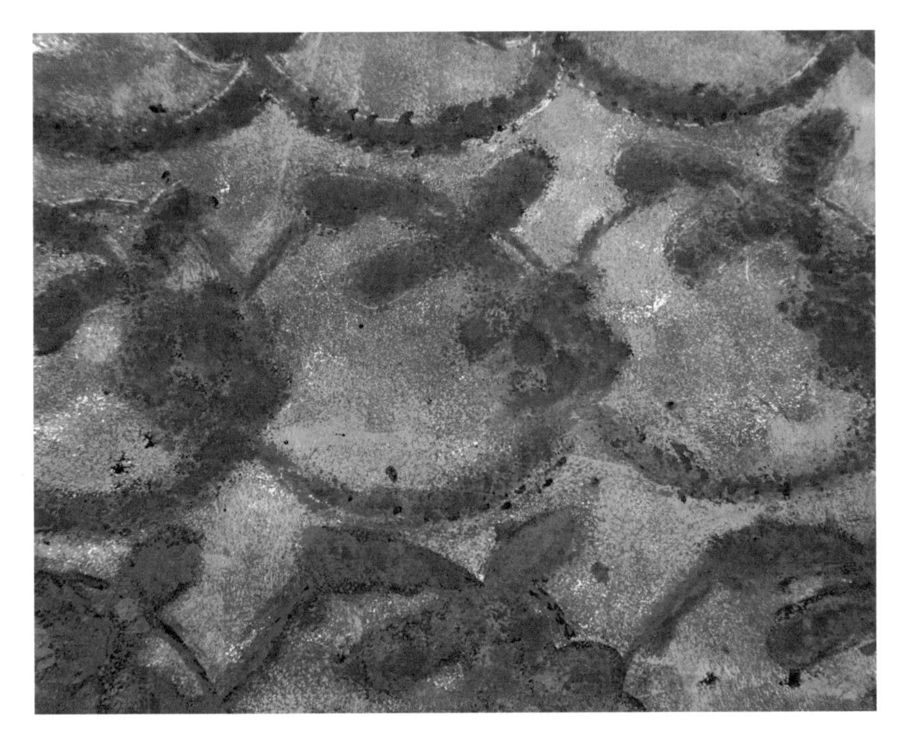

Fig. 19 Fabric project / fine textile realised on tempera on paper; student-author Claudia Davidescu, second year, licence, coordination: PhD Associate Professor Dorina Horătău, 2019–2020. *Photo credit* personal archive Dorina Horătău

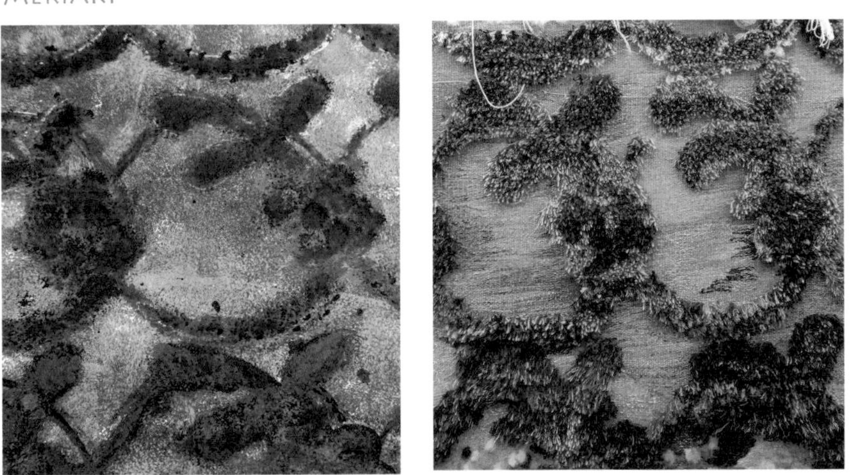

Fig. 20 Upholstery cardboard for tapestry (detail) and haute-lisse completed tapestry woven on Persian knot; student author Claudia Davidescu, second year, licence, coordination: PhD Associate Professor Dorina Horătău, 2019–2020. *Photo credit* personal archive Dorina Horătău

References

1. Hardy A-R (2003) Art Deco Textiles, the French designers. Thames & Hudson Ltd. London
2. Hostiuc C (2008) The Romanian Baroque Answers and Echoes, NOI Media Print Publishing House, Bucharest Gestures of Authority
3. Jones O (2001) The grammar of ornament. The Ivy Press Limited
4. Lorenz K (2007) Civilized Man's Eight Deadly Sins (Romanian translation by Vasile V. Poenaru). Humanitas, Bucharest
5. Meller S, Elffers J (1991) Textile designs, Harry N. Abrams, Inc., Publishers, New York in association with Dohosha Publishing Co., Ltd., Tokyo, DuMont Buchverlag Koln, Ediciones Akal, Madrid, Flamarion, Paris, A. Vallardi, Milano
6. Patterns in Design Art and Architecture, Petra Schmidt, Annette Tietenberg, Ralf Wollheim (eds), Edition form Birkhauser, Basel, Boston, Berlin, 2007
7. Patrimoniu textil/Textile heritage, exhibiton catalogue, Ministerul Culturii si al Cultelor, Muzeul National Bran, Nicoleta Petcu coord., 2007

Webography

8. https://unarte.org/palatebrancovenesti.ro

Handweaving as a Catalyst for Sustainability

Jenny Pinski, Faith Kane, and Mark Evans

Abstract Sustainability has become a key priority within textiles and the associated industries. Whilst there has been a move towards more sustainable practice, there is still some way to go. A multifaceted holistic approach is needed to facilitate meaningful and effective change. Craft practice, including handweaving, has long been regarded as being a more sustainable alternative to mass production, as it promotes the production of high-quality goods and encourages slow rates of consumption. Additionally, handweaving has the potential to contribute via a range of other strategies including, but not limited to, modular design, on-demand localised production, added value through customisation and more. The Chapter firstly presents a review of the role of craft, craft-based design and handweaving in relation to key models of sustainability, and the circular economy is a particular area of focus. It then draws upon a case study of practice, undertaken by the first author, which focussed on the incorporation of the design of end products into the handweaving process. Prior professional design experience in footwear and woven textiles enabled the production of a series of fully fashioned, low-waste handwoven sandal upper designs. Practice was used to support data collection, and qualitative data analysis revealed insights into the opportunities and limitations of the design methods and approach. Key areas of potential for handweaving to contribute to the advancement of sustainable design practice are identified. They are then mapped against key principles of the circular economy, and a number of strategies such are identified which can contribute to the advancement of sustainable design practice. The Chapter focuses on the potentiality of the approach, and future testing is required to further validate the theories presented.

J. Pinski (✉) · M. Evans
Loughborough University, Loughborough, UK
e-mail: J.Pinski@lboro.ac.uk

M. Evans
e-mail: M.A.Evans@lboro.ac.uk

F. Kane
Massey University, Palmerston North, New Zealand
e-mail: F.Kane@massey.ac.nz

M. Á. Gardetti et al. (eds.), *Handloom Sustainability and Culture*, Sustainable Textiles: Production, Processing, Manufacturing & Chemistry,
https://doi.org/10.1007/978-981-16-5665-1_4

Keywords Circular economy · Handweaving · Craft-based design · Sustainability · Responsible design · Zero-waste · Woven textiles · Footwear design

1 Introduction

The global impact of the textile industry on the environment and society has been well documented and discussed. In recent years, sustainable development within this industry has become a priority, with the majority of companies and associated brands having a policy on sustainability [23]. However, whilst the industry is taking steps to become more sustainable, there is still some way to go. Implementing meaningful change that addresses negative impacts on the environment and society whilst maintaining economic sustainability is a complex undertaking.

The environmental and social impacts of large industrial models of consumption have been a key issue within the textiles industry since the Industrial Revolution [4]. Craft practice has been seen as an antidote to mass manufacture, for example, the Arts and Crafts Movement was formed in response to the negative environmental and social impacts of the Industrial Revolution [1]. Handweaving, considered as a craft practice, has a natural alignment to key aspects of current thinking in design for sustainability. It promotes slow rates of production and consumption, and the production of high-quality goods offer longevity which can lessen the impact on the environment by reducing the volume of goods required and minimise waste. However, such products are often inaccessible to most people due to relatively high prices when compared to mass manufactured items. There are also links with handweaving to self-expression, wellbeing and cultural heritage which can contribute to social sustainability. In addition, handweaving has the potential to contribute to furthering sustainable development within the textile industry in ways that go beyond producing high-quality products for more affluent customers. The handweaving process allows time for reflection and opportunities for design innovation, through which in-depth knowledge of materials and novel construction methods can be generated, and innovations such as low- or zero-waste products can be created. This, along with other opportunities, is identified and discussed in this Chapter. In this sense, handweaving is proposed as a tool for design for sustainability. And, through the support of digital and automated processes, the resulting textile-based products may be produced at an accessible price point.

The research presented in this Chapter aims to identify if handweaving has the potential to support and therefore accelerate advancement towards a circular economy, and additionally, which areas of opportunity would benefit from future investigation. The literature review outlines the historical context of the role of craft practice and handweaving in relation to highly industrialised societies. It outlines and discusses opportunities and considerations when using craft processes and principles to support sustainable development. The circular economy is then outlined and discussed in more detail, and examples of the use of handweaving and automated weaving technologies to create low-/zero-waste fully fashioned products presented. Following the literature review, a hands-on case study is introduced, methods outlined and findings discussed. During the case study, handweaving was used as a method of trialling structure and material combinations for a range of innovative sandal designs. In addition to research into material and structure, handweaving was used to develop and refine initial design ideas into final design proposals. The opportunities and considerations identified from the case study are discussed in relation to key principles of the circular economy, and areas for the contribution of a handweaving approach presented. Future research opportunities are identified in testing the areas of opportunity with the aim of identifying unforeseen and negative impacts of the suggested methods and approaches. Additionally, there may be further opportunities for handweaving to contribute to the principle of regenerating natural systems that were beyond the scope of this research.

2 Context

2.1 Craft, Design and Handweaving

Before the Industrial Revolution of the late 1800s, it was common for skilled craftspeople to create one-off products for individual customers [5]. The rise of mass production led to a division of labour and changed the way in which products were designed. Sketching, for example, is a relatively new development that was rarely used by craftspeople [17]. This method of designing, in conjunction with production line manufacturing, limits hands-on experience and distances the designer and production workers from the final product. This can lessen the understanding of materials and constructions which arguably results in less informed decision making.

Pye [28] describes craft as "workmanship of the better sort" (p. 4), but struggles to draw a line where craftsmanship ends and ordinary manufacture begins. However, where risk is involved in the manufacturing process, this is more arguably identified as craft according to Pye. In contrast, Risatti [29] puts forward the theory that craftsmanship is distinguishable from workmanship, and that it encompasses a number of concerns, including the overarching concept associated with a product and the skill in executing it through hands-on making. In terms of drawing a distinction between craft and design, Risatti argues that designers do not conclude their process with a finished product, but a representation of it, to be produced by another person or machine. These examples illustrate the complexity in defining the terms associated with the discussion of handweaving to support and act as a catalyst in design for sustainability. The term "craft-based" is utilised throughout the Chapter to discuss handweaving processes and products that are rooted in craft traditions but are often used as a method of designing. In the case study and some of the examples discussed, the outcomes provide a model for manufacture as opposed to a final product. In this sense, much of the discussion is based around the use of craft-based handweaving as a tool for discovery in order to enhance the designer's ability to make more responsible design decisions.

There is a historical context to consider in the use of craft processes to counteract the impact of mass production. The environmental and social impact of Industrial Revolution sparked the development of the Arts and Crafts Movement (ACM) which used craft-based principles, processes and products to try to counteract the environmental and social damage caused by industrialisation. The principles associated with the Movement commonly refer back to pre-industrialisation where involvement in production was regarded as much more positive for workers. Products were also considered as more beautiful, and there was a strong sense of community that was built through the production of handcrafted products in many different disciplines [1]. However, there were a number of different schools of thought associated with the ACM. For example, key figures such as William Morris and Charles Robert Ashbee completely rejected mass production of any sort. However, the architect Frank Lloyd Wright embraced machine production and the social and creative advantages of it [1]. These contrasting values and divided opinions in the role of the machine in craft, or craft-based processes, may appear to be problematic. However, this Chapter argues the need for a variation in strategies in order to facilitate meaningful progress in terms of the multifaceted challenge of creating a more sustainable future.

The weaving process has evolved a great deal throughout history. The earliest examples of woven textile production consisted of passing weft yarns over and under warp yarns stretched over a frame [4]. Hand looms have since become varied in terms of their complexity and level of automation. During the Industrial Revolution, machine powered weaving technology gained vast momentum [7], and with the development of the power loom, manufacturers were able to produce quality cloth quickly and without the need for highly skilled hand weavers [4]. Despite increased use of machine power and the opportunities afforded with automated production, weaving by hand is still a relevant practice as it provides greater flexibility than that experienced on mechanised looms [2]. This supports the relevance of the use of handweaving within both design for scaled-up production as well as small-scale production undertaken by individual makers or artisans. The role of handweavers and craftspeople has evolved over time, and Lommerse et al. [19] found that when an awareness of commercial production processes is developed, it is possible to use this knowledge to reach a wider market. However, artisans would need to consider the implications for sustainability in order to utilise commercial production processes in a responsible way.

It is not only from a production point of view that technology can support craftspeople. Online marketplace platforms, such as Etsy, have provided wide audiences for handmade products and aim to build communities of makers. Buyers can shop locally using the platform or access craft products with global reach. The reality is often that small business or individuals selling on such platforms end up implementing a division of labour and/or working long hours for little pay in order to be competitive [16]. Crafts collectives and galleries, which are positioned at the high end of the market and are selective in the craftspeople they represent, are also making use of online commerce to reach a global audience. An example is the The New

Craftsmen, a UK-based gallery and online platform, which sells carefully selected, high-quality crafts products [24]. The makers represented through this platform are of the highest standing and, due to a higher price point being achievable, there is not the same requirement for competitiveness by emulating aspects of mass production or lowering rates of pay. The high cost of the products means that they are not accessible to the majority. Through these two examples, it is possible to see the complexity and difficulties craft practice has in maintaining its social and environmental integrity.

2.2 An Arts and Crafts Model for Sustainability

There are a number of differing approaches to developing more sustainable practice within the textiles industry. One of the most established approaches aligns with the aims of the ACM, as it focuses on the design of high-quality pieces that enable slow consumption by promoting longevity. Whilst this is undeniably good practice in design for environmental sustainability, the resulting products are often of high cost and not accessible to the lower end of the market, which limits the reach of the approach. By example, the ACM ultimately went into decline, because the resulting products became largely exclusive to an elite market due to the high cost of production. Although the ACM aimed to be inclusive, it ultimately proved to be unattainable [1]. Although a slow approach such as that which was championed by the ACM has limitations in accessibility, many aspects of it are inherently sustainable and without its existence many skills and traditions could be lost. Therefore, there is a need and opportunity for craft-based skills, traditions and logic to contribute to sustainability beyond the production of long-lasting products made by a skilled individual maker.

2.3 Handweaving and the Circular Economy

A more recent development that has become a well-established concept within textile design and beyond is the strive towards creating a circular economy. This involves three key principles—design out waste and pollution, keep products and materials in use and regenerate natural systems [12]. Unlike the ACM model, it is highly a flexible and holistic approach towards more sustainable production and consumption. It encompasses frameworks such as Cradle to Cradle, which is concerned with the flow of materials and eliminating waste by considering it as a source of nutrients or energy that can be fed back into the cycle. This particular school of thought was popularised by the work of Braungart and McDonough [6] and it is now a well-known strategy. Whilst there has been some positive developments within the textiles and associated industries, there is some way to go in terms of implementation of circular rather than linear models. This is acknowledged within the industry. For example, US fashion label Eileen Fisher aims to embody a circular design approach. However, whilst progress has been made, the company acknowledge the need to "go further, faster" with their development in this area [10]. In relation to this concept and in acknowledgement of the need for flexibility and differing approaches within the textiles industry, Goldsworthy et al. [15] present a case for the need for different cycle speeds within a circular approach. They consider it as being important to consider the appropriate speed for individual products and sections of the market as opposed to enforcing the idea that slow living is the only path to a more sustainable future. This means that different sustainable strategies become more or less important depending on the speed of the cycle and in terms of material usage. For example, Goldsworthy et al. [15] describe the way in which recent technological advances in material recovery are opening-up opportunities to recycle materials to production quality using "fibre-to-fibre recycling technologies" (p. 62). The use of such materials may become more vital for use in products with a relatively short lifespan as opposed to those which are likely to remain in use for decades or even generations.

The development of new more sustainable material options is one strategy that could enable designers to make more responsible choices. Hands-on experience with these materials can allow designers to explore new opportunities and evaluate the suitability for use in a particular product. A strong relationship in the understanding of materials can be gained through hands-on interaction [8, 18]. Through handweaving, the maker touches every warp thread when setting up a loom. They carefully interlace the yarns and gain embodied and tacit knowledge of how a material behaves in a highly informed approach.

Piper and Townsend [27] present an approach for creating whole garments through both handweaving and digital Jacquard processes. The handweaving element was found to support digital design by providing embodied knowledge of materials and processes. When considering the end product plus material and construction in parallel, it is possible to make more informed decisions and provide greater control with in-depth knowledge of the product and production methods [26]. It is also possible to apply appropriate yarns to weave structures to provide optimum results in producing the required properties for a particular product. The design outcomes may remain as one-off pieces or they can be adapted and scaled up to create more affordable products, for example, through the use of automated manufacturing. Piper and Townsend [27] described the way in which handweaving can inform the transition to digital methods of design and production. Within the context of whole garment weaving, Piper developed a collection of fully fashioned garments using a Jacquard power loom. Similarly, McQuillan [21] has developed a range of multi-layer woven garments using an automated Jacquard loom and has identified the potential for this method of production to enable on-demand, localised production on a human scale. At the bespoke end of the market, multi-layer handweaving is being used to create one-off sculptural pieces. Azure [3] creates sculptural installations in an approach that stems from a background in jewellery, metal arts and handweaving. Whilst the sculptural pieces created are high-cost one-off art pieces, many are displayed in public spaces and are therefore available for a wide audience to view and enjoy. These examples demonstrate different ways in which artists, designers and researchers are creating artefacts using weaving as a method of generating three-dimensional form and how this can inform sustainable design practices.

3 Hands-On Case Study

This section reports on a case study of practice undertaken by the first author who, before becoming an academic, spent 4 years as a practicing footwear designer. The project incorporated handweaving in order to design a collection of fully fashioned low-/zero-waste sandal designs. Handweaving was supported by digital design technologies in order to successfully visualise final design proposals. The case study analysis aimed to identify if handweaving has the potential to support and therefore accelerate advancement towards a circular economy. Additionally, areas of opportunity for future research were determined.

3.1 Methods

A brief was created for the project that outlined the research aims, objectives and design information. It contained written detail of the research rationale, market, competition and three mood boards that explored woven structure, colour and outsole inspiration. The outcomes consisted of four upper designs that were displayed on top of 3D printed outsoles (see Fig. 1). The designs were used as part of a focus group with established footwear design professionals and were presented at an international academic conference, Textile Intersections, 2017.

The project took place over a period of 20.5 weeks and design activity was carried out on 64 separate days with an average of 4.2 h of designing per day. The woven structures explored within the project were multiple-cloths and block structures which

Fig. 1 Three of the final design proposals created during the case study

Fig. 2 A flat upper showing an example of the multiple straps achievable through multiple cloth and block constructions

built on existing work by the first author. A zero-waste, stitch-free concept was explored, and multiple sandal straps were created within a single piece of cloth. The set-up of the loom allowed the researcher to split the cloth into multiple layers and also into narrower or wider straps. It could be woven as one large strap, or it could be woven as numerous separate pieces (see Fig. 2). The yarn used was a nylon cord for stability combined with a matt viscose to add softness against the foot.

The design process was recorded and reflected on in an approach aligned with Frayling's "research through art and design" [14] (p. 5). In addition to documenting the process, the design outcomes were also treated as data. An overview of the data formats is presented below along with the amount of data collected:

- Diary sheets × 20
- Diary log entries × 64
- Design sheets × 11
- Digital files × 16
- Physical artefacts or sample sheets × 39

The documentation methods were based on those detailed by Pedgley [25] and consisted of a diary log and diary sheets. Diary writing was used as a method of reflection and allowed "self-conversation" [25] (p. 472) in response to a day of designing. An overview of design activity was kept via the diary log which kept track of all design activity and linked it with the corresponding file and/or artefact. Additionally, visuals with notes were used to reflect on samples once they had been taken off the loom. This aided the researcher in making design decisions through critical judgements (see Fig. 3). The sheets were printed and, where possible, kept

Weave as a method of sandal design:
Case study III "digital integration" - Sample notes

Warp number: 4

Sample number: 1

Notes

Fabric composition: Approx 70% nylon, 30% viscose

Finishing:
Press

Observations/comments:
The proportions are not correct for this upper design, a mistake was made when weaving. This needs to be sorted out for the final sample to be the same as warp 3 sample 2. The pink strap is too flimsy, this would benefit from having some nylon in the warp and multiple viscose ends wound as one. The folds and joins are a bit distorted but I think this is because the proportions are not correct.

Changes for next sample:
- Pink strap to be more substantial
- Proportions to be as per warp 3 sample 2

Fig. 3 Sample analysis sheet

in a folder with the actual samples for reference. This was an effective method of keeping track of and analysing woven samples in a systematic way, the process was valuable to design development and the resulting sample sheets filed with physical samples aided data analysis.

3.2 Analysis

Data was analysed using qualitative methods informed by Dey, Miles and Huberman and Eisenhardt [9, 22, 11]. The process consisted of the completion of a series of steps, data reduction, clustering, coding and display. The variety of data collected within the case study meant that information from a number of sources was consolidated and focussed at the data reduction stage. It was split by activity/task, and NVivo software was used to group and cluster the relevant information. The first stage of analysis was to transcribe the hand-written data within the diary entries and log. This translated it into an appropriate format for analysis using digital software and re-familiarised the researcher with the data. It was then reduced/re-organised into tasks, and each one was coded according to the stage of the design process, the approach used and whether it informed the final designs. This grouped the data into a form that enabled the researcher to gain an overview of the design process and its structure. Following the coding process, it was possible for the researcher to consider the data in its re-structured format. Statistical analysis revealed how many of the handweaving tasks undertaken were deemed to be successful in informing the final design outcomes and this enabled initial patterns to be identified. These were further explored and discussed in relation to the relevant descriptions, insights and design outcomes.

3.3 Findings

Figure 4 shows an overview of the design process and approaches used in the case study project. As the arrows on the diagram indicate, the process was not linear and stages returned to throughout. The design stages were based on Wilson's [31] general design process model and they were developed in relation to the data. Although designers do not generally follow a formal model, it is possible to identify common elements [20] and this structure aided analysis. The stages consisted of concept development, initial research, in-depth research, design development and presentation.

Concept development is the stage before the project brief has been written. It acknowledges that the researcher processed ideas and previous research and development in order to define the concept and consider elements such as identifying appropriate woven structures. The initial research stage consisted of writing the design brief, creating mood boards and sourcing yarn to begin weaving. Idea generation was the point where initial design ideas were generated by drawing and quick

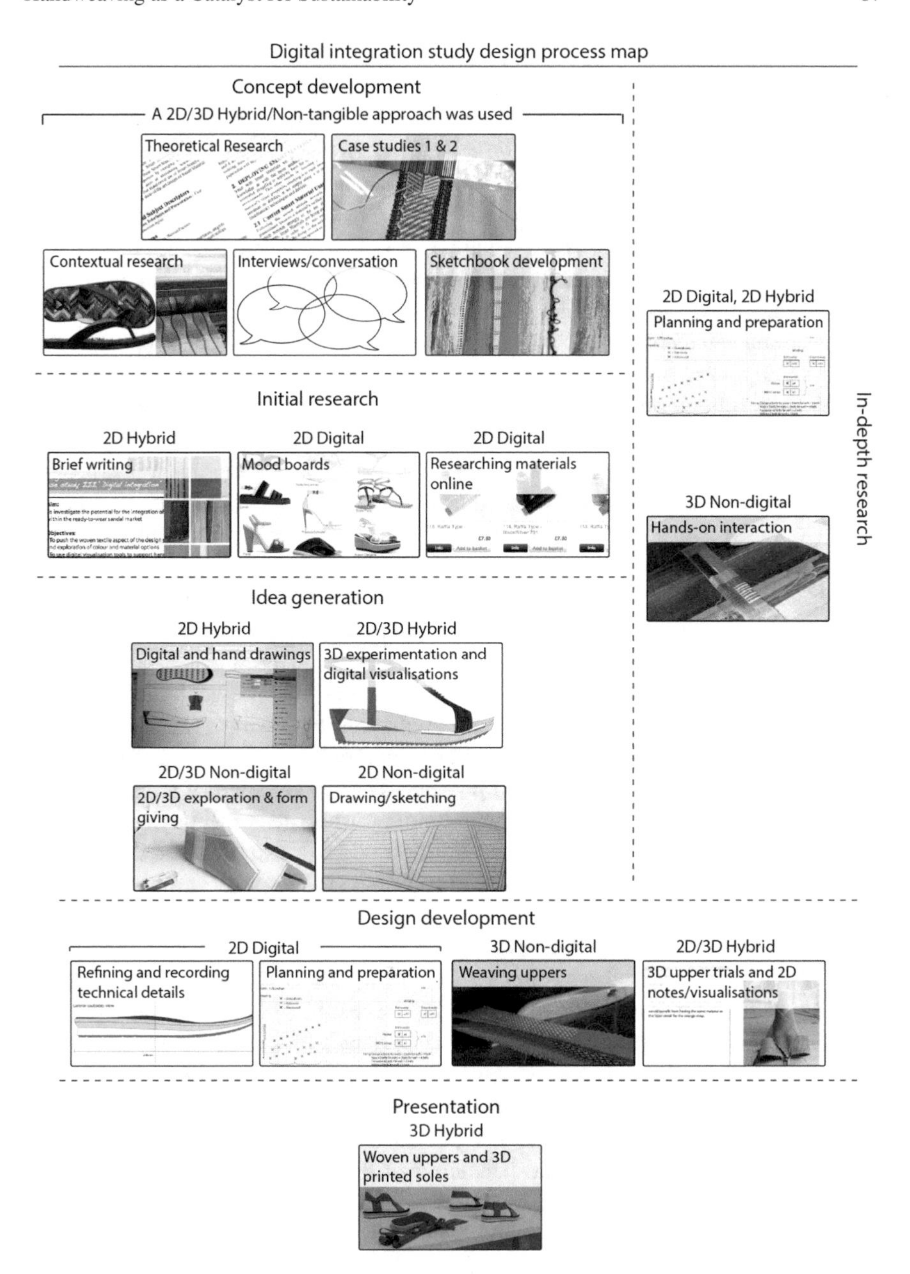

Fig. 4 Diagrammatical overview of the design process and approaches used

methods of 3D modelling by hand. In-depth research consisted of handweaving to trial and explore different material and structure combinations. Design development is the point where design ideas were woven in full and the patterns, structures and materials refined through a sampling process. At this stage, the outsole design was finalised, and a specification sheet with measurements and cross sections sent to an expert in 3D printed models. Finally, presentation was where the 3D printed outsoles were brought together with the woven uppers to show a finished design proposal.

Handweaving was used for in-depth research and design development. Therefore, the discussion of the findings now focuses on these two stages in relation to handweaving and its potential to contribute to the move towards a circular economy. At the in-depth research stage, handweaving was used to trial ideas for structures, material and colour combinations. In a previous project, the researcher had explored the use of a medium weight (2/6 s) cotton yarn which did not provide enough structural integrity for the purpose of footwear. It was also deemed to not fit the desired aesthetic which had evolved to be inspired by sportswear, and the cotton provided a casual look. In response to this, a nylon cord was used, and this was found to have the benefit of shape-holding properties when woven. However, the resulting material had a rough surface and therefore would not be comfortable for wear against the foot. For this project, the researcher firstly selected a matt viscose yarn that had a soft finish, with the potential to be structurally sound through the integration of more substantial weft yarns, and it fitted with the overall aesthetic of the project which was now more focussed toward synthetics and semi-synthetics. Following a range of material trials, the viscose and nylon yarns were both incorporated, and this was found to be the most appropriate for the intended product. The combination of softness of the viscose alongside the strength and shape-holding properties of the nylon cord was found to be effective. The properties of comfort and strength achieved through the material combination have the potential to contribute to longevity at the use stage of the product lifecycle.

The focus of the project was to create a fully fashioned, stitch-free range of footwear, and the material selection was based on the aesthetic and performance qualities which may contribute to extending the useful life of the resulting products. However, consideration of aspects, such as recyclability and the environmental impact of yarn production, would have also been beneficial at this early stage. Trialling materials through handweaving allowed the researcher to test and evaluate a variety of potential yarns. It also provided the opportunity to gain in-depth knowledge of the handle, feel and behaviour of the materials. This eventually led to the combination of yarns that were tailored to be best suited for the intended outcome. This approach would allow a handweaver to explore materials that align to a range of sustainable design strategies such as those associated with different circular speeds as outlined by Goldsworthy et al. [15]. This type of in-depth investigation enables designers to adapt and problem solve to develop novel material and structure combinations. Through this experimental approach, there is potential to deliver novel results.

Data analysis of the in-depth research stage of the design process revealed a key opportunity in the capacity for handweaving to facilitate the generation of knowledge of materials and constructions. It revealed a number of reflections and remarks on the levels of effectiveness of techniques and materials that had been trialled. For example, in the design diary, the researcher noted, "I tried weaving the three warps separately, this seemed slightly less stable with some more broken ends appearing than had been before, particularly on the viscose warp". In this case, the amount of nylon was increased to improve the structural integrity, and the results were more successful.

A key theme to emerge from the data was the large number of problems encountered with the woven samples. Knowledge was gained through a trial and error approach, and learning from unsuccessful outcomes became important in facilitating knowledge generation. Another relatively prominent theme was the intuitive and explorative nature of the process, as data revealed that the researcher was reflecting and making design decisions throughout. For example, the researcher commented, "the graphic detail worked well and I may include this in a design—it is more interesting than plain weave and goes well with the look of the project". This demonstrates how the in-depth research was not purely a technical task in testing and developing novel materials and constructions—it involved a substantial amount of aesthetic development in parallel. The design outcomes at this stage varied in quality, and one warp in particular, which explored the use of a fan reed, was problematic. The weave structures had unintended areas of different warp densities which meant the quality was not to a presentable standard. In another warp, the materials were too fine to be suitable, but the use of multiple-cloths and block constructions worked well to create multiple sections which were translated into multiple strap sandal designs. Exploring unsuccessful avenues through the weaving of both warps, informed aspects of the final designs, as the successful elements were pushed forwards and unsuccessful ones were abandoned. Therefore, though it was time consuming and uncertain, it was a necessary part of the process. Handweaving fed into and informed the generation of design ideas and it opened up new possibilities through trial and error. However, the suitability of the outcomes may not be immediately evident until they are refined through further development.

Handweaving was also the primary method used for design development, the stage at which design ideas are sampled and refined. Design development was the most time-consuming stage of the project and accounted for 47% of the overall time spent. In contrast to in-depth research which was highly open and uncertain, this stage had a strong focus towards development of final designs. Though time consuming, it was a focussed and productive method of generating final design outcomes. Digital design software was used to visualise design ideas which at this point consisted of a flat drawing of an outsole and handwoven samples (see Fig. 5). In relation to this, the researcher noted, "I was also able to put an upper design onto the sole in order to visualise how it would look. This was useful and will help in selecting a design to proceed with". Visualisation aided the decision-making process and, in addition to bringing different formats of design elements together, it was used to combine and try variations of outsole designs that were drawn up digitally. The researcher noted, "I was able to take aspects of one sole design and apply them to another". In this respect, it was an efficient way of working, due to the ability to amend and merge designs

Fig. 5 A digital visualisation of a design idea

with ease. Technical details were also developed and recorded using digital software, this aided decisions through the introduction of technical parameters at key junctures such as planning a new warp in order to progress the project. In addition to aiding decisions, the drafting activity involved the generation of knowledge of constructions. This particular pattern was unexpected due to the representational and abstract nature of the method. However, the data revealed that the generation of technical details was useful in clarifying and developing knowledge of the structures. Knowledge gained from previous sampling was adapted and used to draft new warps and create lifting plans. In terms of efficiency, a digital approach enabled the researcher to amend existing plans as opposed to starting from scratch.

The majority of design activity consisted of handweaving, and prominent themes found within the data were "generating knowledge of constructions" and "successful problem solving". Much of the knowledge gain and problem solving was done via the evaluation of samples once taken off the loom. The researcher also gained construction knowledge whilst weaving and was able to solve problems/issues on the loom. For example, when weaving upper 4, a design that incorporated a pleated structure, the researcher came across an unexpected issue with the construction that was solved on the loom. The researcher described this in the design diary, "I had to unpick the pleat and have a re-think. I got around it by weaving the top warp at the same time and when I wove in the pleat I took the tension off the top and middle warps. This allowed for the volume of the pleat between the layers". The issue related to one of the pleats having no space to fold into. This was only brought to light through weaving and the solution was not immediately obvious. Having a deep knowledge of the problem and construction was key in solving this issue and the design was taken forward as a final piece (see Fig. 6).

In some instances, weaving was regarded as a means of producing a design as opposed to developing it, "I have nearly finished weaving upper 1, it seems to be going well with no unexpected problems so far. It has seemed like more of a practical task, following plans rather than letting it evolve on the loom". This was at a point when there were no issues with the construction.

When setting up the loom, the researcher identified and solved problems via reflection using tacit knowledge, "I realised something was not quite right as I put the pegs in the lag belt. It gave me time to think as I was putting them in. It seemed something was not quite right and I was correct". All of the problem-solving activity

Fig. 6 Pleated detail from one of the final upper designs

was deemed to be successful, and upon reviewing the data relating to issues with the outcomes, these were all solved and not abandoned as was sometimes the case at the in-depth research stage. The theme of uncertainty was more prominent in the data relating to the design development stage in comparison to in-depth research. This was unexpected due to the focussed nature of the process in comparison to the more open and explorative approach used for in-depth research. When reviewing the data in relation to this, the main cause for uncertainty was the lack of ability to see whether the designs were successful when still on the loom. They needed to be cut of and analysed regularly. Therefore, while construction knowledge was gained on the loom, enabling problem solving such as the pleating example, material development required an iterative approach that involved more uncertainty and less immediacy in the ability to solve problems.

4 Discussion

The design outcomes of the case study consisted of a range of sandal upper designs that were woven to the shape with the correct strap widths and configurations built into the woven construction. Therefore, there is no need for cutting, stitching and/or gluing which is usually associated with textile footwear production. This demonstrates the ability to use handweaving to generate novel low-waste construction methods in sandal design. Therefore, handweaving can contribute to design intervention that can reduce production waste. This is echoed in the practice and research of Piper and Townsend [27], where multiple cloth constructions were used to generate fully fashioned garment designs. They were woven to shape on the loom with no need for cutting and stitching and therefore cutting waste is eliminated. The approach also has the ability to span beyond fashion and footwear, and more structural three-dimensional forms are also possible. The work of Azure [3] is an example of an artist/designer who creates sculptural pieces using similar methods and approaches. A key aspect of Azure's work is the integration of 3D form into the woven structure and her work spans across the metalwork and weaving. The artist creates sculptures on the loom which are formed into 3D pieces, and like the work presented in the case study, the pieces are woven to shape, and as a consequence, the amount of waste material generated is minimal. In addition to the use of low-waste fully fashioned weaving processes, Azure approaches sustainability from other angles by supporting communities and the environment. This is done through involvement in three different schemes, Mayan Hands, 1% For the Planet and Berkley's Monofilament Recycling Program. Through the Mayan Hands scheme, Azure has been able to teach new techniques to Mayan women to provide opportunity for them to expand their product ranges and support them in generating income through craft practice [30]. This demonstrates how furthering craft practices and developing novel methods of handweaving has the power to benefit the environment through lowering waste, it can benefit communities through knowledge sharing, and through schemes such as 1% For the Planet, businesses and craftspeople can give back and make progress in other ways.

The case study findings indicate that there is value in using a handweaving approach to support design for sustainability. It can provide opportunities to reduce waste through the design of textiles and products in parallel using a fully fashioned approach. The findings are now discussed in relation to the three key principles of the circular economy, "design out waste and pollution", "keep products and materials in use" and "regenerate natural systems" [12].

The craft-based woven textile approach investigated within the case study lends itself to addressing the first two of these principles. Low-waste methods of construction are a starting point in eliminating waste. However, there is likely to be some production waste at the beginning and end of the warp, and this, in addition to post-consumer waste, needs to be considered and eliminated. In terms of production waste, the ability to recycle excess yarn is one potential approach, as the approach used by Azure and their involvement in Berkley's Monofilament Recycling Program demonstrates. Another strategy is that the waste has the potential to be recycled back into handwoven products. The flexibility afforded with the use of handweaving means that it is possible to take waste yarn and weave it back into new products. There will always be some waste from the warp, however, if this can be incorporated into new, useful products, then this is another way to approach the issue. Handweaving can enable such an approach and make it viable. In terms of post-consumer waste, there is potential to address this in a number of ways. Following the case study, the final design proposals were presented to a group of footwear design professionals, and the potential for the sandal designs to be sold as a modular product was identified. With some further development, the upper and outsole could be designed as such that the upper section could be removed and customised by the wearer, the input of the consumer into the final product prolong the life of the product due to the added value and sense of ownership that can be created through participatory design [13]. This in turn helps to keep products and materials in use. However, eventually the upper and the sole will either need to be repaired or disposed of once they have come to the end of their useful life. It is important for designers to consider what happens at this point, and there are different ways in which handweaving can contribute to good practice.

The case study found that, firstly, through a process of in-depth research through handweaving, different combinations of yarn and structure can be explored and trialled. The in-depth knowledge gained of the materials enabled the researcher to develop a combination that capitalised on the shape-holding properties of one yarn and the softness of another. This facilitated the creation something that was fit for purpose and aesthetically pleasing. This type of approach would also be suitable in taking a selection of materials chosen for their sustainable properties and developing yarn and structure combinations that are applicable for new applications. Additionally, the ability to generate design proposals that consist of multi-layer woven constructions eliminates the need for harmful substances such as adhesives to be used to hold together different layers of the products. Instead, the structure holds its multiple layers together. This not only helps with the principle of designing out pollution, it also means that the uppers can be taken apart once they have reached the end of the use phase of the cycle, and this opens up potential for yarn types to be separated and recycled, or potentially re-woven into another product. Strategies in which materials from post-consumer waste are reconstructed into new products are relevant and applicable to craft-based design and production [15]. The flexibility associated with handweaving means that there is also potential for a customisable on-demand system. This may consist of creating bespoke colourways using different weft yarn colours and adapting patterns and styles with ease. Whilst there are limitations associated with the loom's set-up, there is scope to explore different structure and material combinations on the same warp. This can improve desirability and add value and longevity through creating a sense of ownership and attachment to the product [13]. The ability to produce products to order can help reduce unnecessary waste, and this approach to production has the potential to be localised which has the added benefit of reducing transport emissions [15].

Whilst the case study was able to help highlight potential in relation to two of the three principles of a circular economy, the third, to "regenerate natural systems", did not return any obvious findings. It was beyond the scope of the study to analyse systems through this lens. However, this is an area where future research would be valuable. For example, waste products, such as water used in fibre production and yarn finishing or dyeing, may be able to be recirculated back into natural systems and provide nutrients.

5 Conclusion

There are a number of ways in which handweaving can support design for sustainability, and in particular, the use of handweaving as a method of designing materials and products in parallel has potential to facilitate change. Designing for a circular economy is a complex and multifaceted process where a depth in understanding of every element of the product and the whole lifecycle is vital. The in-depth knowledge generated through handweaving has potential to give designers a more all-encompassing knowledge of the product and production methods. This can facilitate the ability to make more informed decisions regarding different aspects of the product's lifecycle. Table 1 shows the areas of potential that were identified and mapped against the corresponding principle of the circular economy.

The discussion around the case study provides initial insights into the potential for handweaving to facilitate and accelerate the advancement towards a circular economy. As shown in Table 1, the key areas of potential which were identified map onto two of the three principles that have been outlined. Therefore, investigation into the potential for handweaving to contribute to the third principle of circular design which is concerned with regenerating natural systems is also an area of opportunity for future research. When implementing new design and business strategies, it is possible that there may be unexpected consequences, and therefore, any system and approach that is implemented successfully must be carefully tested, implemented and monitored to reduce negative impact. This is an area which also requires future investigation.

Table 1 Areas of potential for handweaving to contribute towards sustainability mapped against the corresponding principles of the circular economy

Area of potential	Corresponding principle(s) of the circular economy
Low-waste construction through methods of fully fashioned weaving	Design out waste and pollution
The ability to experiment and test emerging materials that have sustainable properties	Design out waste and pollution
Potential for on-demand production and customisation	Design out waste and pollution; Keep products and materials in use
Potential for development of modular products	Design out waste and pollution; Keep products and materials in use
The elimination of harmful adhesives to reduce pollution and enable disassembly for recycling or re-making	Design out waste and pollution; Keep products and materials in use

References

1. Adams S (1987) The arts and crafts movement. Grange Books London
2. Albers A (1965) On weaving. Studio Vista London
3. Anastasiaazure.com (n.d.) Anastasia Azure: About. https://www.anastasiaazure.com/about. Accessed 16 Jan 2021
4. Benson A, Warburton N (1986) Looms and weaving. Shire Publications Princes Risborough
5. Boër CR, Dulio S, Jovane F (2004) Shoe design and manufacturing. Int J Comput Integr Manuf 17(7):577–582
6. Braungart M, McDonough B (2002) Cradle to cradle: re-making the way we make things. North Point Press New York
7. Buchanan DR (1995) Manufacturing Innovation, Automation and Robotics in the Fibre, Textile, and Apparel Industries. In: Berkstresser GA, Grady PL (eds) Automation in the textile industry: from fibres to apparel. Textile Institute, Manchester, pp 3–15
8. Cross N (2001) Designerly ways of knowing: design discipline versus design science. Des Issues 17(3):49–55
9. Dey I (1993) Qualitative data analysis, a user-friendly guide for social scientists. Routledge, London
10. EileenFisher.com (n.d.) Horizon 2030: What if we did things differently. https://www.eileenfisher.com/horizon2030. Accessed 18 Jan 2021
11. Eisenhardt KM (2002) Building theories from case study research. In: Huberman M, Miles MB (eds) The qualitative researchers companion. Sage London, pp 5–35
12. Ellen MacCarthur Foundation (n.d.) *Concept: What is a circular economy?* A framework for an economy that is restorative and regenerative by design. https://www.ellenmacarthurfoundation.org/circular-economy/concept. . Accessed 7 Jan 2021
13. Fletcher K (2014) Sustainable fashion and textiles: design journeys 2. Routledge, London
14. Frayling C (1993) Research in art and design. R College Art Res Papers 1(1):1–5
15. Goldsworthy K, Earley R, Politowicz K (2018) Circular speeds: a review of fast and slow sustainable design approaches for fashion and textile applications. J Text Des Res Practice 6(1):42–65
16. Krugh M (2014) Joy in labour: the politicization of craft from the arts and crafts movement to Etsy. Can Rev Am Stud 44(2): 281–301
17. Lawson B, Loke SM (1997) Computers, words and pictures. Des Stud 18(2):171–183
18. Leader E (2010) Materializing craft: evaluating the effects of experiencing actual materials during the design process. Des Principles Practices 4(4):405–418
19. Lommerse M, Eggleston R, Brankovic K (2011) Designing futures: a model for innovation, growth and sustainability of the craft and design industry. Des Principles Practices 5(4): 385–404
20. McDonagh-Philp D, Lebbon C (2000) The emotional domain in product design. Des J 3(1):31–43
21. McQuillan H (2019) Hybrid zero waste design practices. Zero waste pattern cutting for composite garment weaving and its implications. In: Proceedings of EAD 2019, running with scissors, 13th EAD conference, University of Dundee, Nethergate
22. Miles MB, Huberman AM (1994) Qualitative data analysis: an expanded sourcebook. Sage London
23. Muthu SS (2020) Assessing the environmental impact of textiles and the clothing supply chain 2. Woodhead Publishing Duxford
24. Newcraftsmen.com (n.d.) The New Craftsmen: About Us. https://www.thenewcraftsmen.com/about-us/. Accessed 18 Jan 2021
25. Pedgley O (2007) Capturing and analysing own design activity. Des Stud 28(5):463–483
26. Pinski J, Kane F, Evans, M (2018) Craft-based design for innovation: potential in novelty, quality and sustainability through hands-on interaction. Artifact: J Design Practice 5(2):3.1–3.20

27. Piper A, Townsend K (2016) Crafting the Composite Garment: The role of hand weaving in digital creation. J Text Des Res Practice 3(1–2):3–26
28. Pye D (1968) The nature and art of workmanship. Cambridge University Press London
29. Risatti HA (2007) A theory of craft: function and aesthetic expression. University of North Carolina Press Chapel Hill
30. Rosenbaum B (2019) Crafting Change. https://www.mayanhands.org/blogs/news/title. Accessed 16 Jan 2021
31. Wilson JA (2011) Core Design Aspects. PhD thesis, The University of Manchester

Handloom—The Challenges and Opportunities

Sankar Roy Maulik

Abstract Handloom weaving has been practised in different parts of the country, and it represents the vibrant cultural identity of India. The handloom industry is primarily a rural-based activity and one of the largest economic activities after agriculture. The handloom weavers are keeping this traditional heritage alive, although they belong to the most vulnerable and weaker section of society. Handloom sectors provide employment to weavers and allied workers and playing a unique role in the Indian economy. The total number of handloom worker households in India is 31.44 lakhs, and the average number of weavers per household is 1.05. This sector contributes to the export earning of the country and almost 95% of the hand-woven fabrics are exported to the entire world from India. The adoption of new technology, availability of cheaper power loom fabrics, changes in fashion and consumer preference, alternate employment opportunities for better earning and economic liberalization have made serious inroads into this traditional vibrant sector. The prime advantage of this sector is its ability to introduce a variety of aesthetic innovations through experimentation apart from transferring skills from one generation to another. This industry has widened the entrepreneurial base and also has the adaptability to fulfil the supplier's requirement, etc. Environment consciousness and green legislations globally along with consumer taste have made the handloom fabric in demand. Nowadays, people are looking for sustainable products. Hence, new business models with innovative design, materials and processes are very important for the survival of our rich cultural ethos. It is very much essential to develop eco-friendly value-added handloom products to enrich the aesthetic appeal without compromising the product quality. Different eco-friendly dyes including natural colour and varieties of yarns made from natural fibres can be used for producing the value-added products. This value-added handloom product may improve the livelihood of the artisans associated with handloom weaving.

Keywords Dyeing · Eco-friendly · Handloom · Heritage · Natural · Printing · Sustainable · Value-added

S. Roy Maulik (✉)
Department of Silpa-Sadana, Visva-Bharati (A Central University), Sriniketan, India
e-mail: sankar.raymaulik@visva-bharati.ac.in

M. Á. Gardetti et al. (eds.), *Handloom Sustainability and Culture*, Sustainable Textiles: Production, Processing, Manufacturing & Chemistry,
https://doi.org/10.1007/978-981-16-5665-1_5

1 Introduction

The art of weaving is an age-old tradition in India that has been practised over centuries with the earliest evidence going back to the Indus Valley civilization. The export of handloom products was reported as early as the fifteenth century, and the design of handloom fabrics was greatly influenced by geographic, religious and social customs of a particular region. This sector represents the continuity of the Indian heritage of hand weaving and reflects the socio-cultural tradition of the weaving community. Handloom textiles constitute a timeless facet and form a precious part of generational legacy. It also exemplifies the richness and diversity of our country. Handloom weaving constitutes one of the richest and most vibrant aspects to ensure sustenance of our rich cultural heritage. It is one of the largest unorganized economic activities, which constitute an integral part of rural and urban livelihood. This sector was a nationalist activity and became symbolic of the Indian independence struggle. The Indian handloom industry plays a strategic role in restructuring and transformation of the economy and is a reflection of our excellent craftsmanship. It is one of the largest economic activities after agriculture. This heritage industry has the potential to create ample opportunities for employment generation among the rural people. Handloom industry provides green livelihood opportunities to millions of families and supplementing incomes in seasons of agrarian distress. The relevance of the handloom sector in the agrarian economy is massive, uses agricultural products as raw materials, and provides an ever-ready market for agricultural produce. The world is concentrating on green environment concept to save the planet and handloom could be one of the alternatives to fulfil the basic needs of human, i.e., clothing. This sector meets every need ranging from exquisite fabrics, which takes months to weave, to popular items of mass production for daily use. This sector provides employment for the women labour force in family environment and laid the foundation stone of women's leadership, and thus directly addressing women's empowerment. At an aggregate level, 72% of handloom weavers and allied workers are female, i.e., approximately 23 lakhs. In the rural areas, female weavers are more as compared with male counterparts, whereas in urban areas, this ratio is nearly even. This sector contributes nearly 15% of the total cloth production and is also responsible for the export earning of the country. 95% of the world's handwoven fabric comes from India [1]. Handloom is known for flexibility, versatility and innovativeness. The level of artistry and intricacy achieved in the handloom fabric is unparalleled and certain weaves/designs are well beyond the scope of modern weaving machineries. The handloom industry has deep roots in the cultural diversity that distinguishes India from rest of the world [2]. The people engaged in this profession are mainly from the vulnerable and weaker section of the society, but they are keeping alive the traditional craft of different states.

The strength of this heritage industry lies in its openness to innovation along with flexibility of production, uniqueness, adaptability to suit suppliers' requirement, etc. This sector can also produce goods in small volumes with exquisite design. Handloom is unparalleled in its flexibility and versatility, blending of myths, faiths, symbols, permitting experimentation, encouraging innovations and generates production at low capital cost. This industry mostly using indigenous raw materials by utilizing local resources, widens entrepreneurial base, facilitates balanced regional growth along with prevention of the migration of labour to the metropolitan areas [3]. However, in the present context of globalization and rapid technological changes, this sector faces manifold challenges mainly due to obsolete technologies, weak supply chain, low productivity, inadequate working capital, conventional product range, weak marketing link, overall stagnation of production and sales, capitalist control, low wages, increased yarn price, etc. The production system is mostly controlled by a particular entrepreneurial class, i.e., master weaver. This sector has the tradition of transferring skills from one generation to another. But, the share of weavers in the age group between 14 and 18 years is only 2.4% [4], which indicates that the younger generation is not keen enough to pursue their profession in this traditional process of making cloth. The challenges and issues of concern to this industry are discussed below.

Rising input costs

The sharp increase in price of yarns, dyes, chemicals and other items have resulted into cost disadvantages to the handloom weavers. This problem is more acute for the individual weaving households who need small quantities of yarn and chemicals.

Credit problem

It is difficult to obtain credit from institutional sources due to the poor financial condition of handloom weavers and allied workers, and they have to depend on the mercy of private money lenders. Hence, the exploitation by the private money lender continues. The lack of rudimentary financial literacy further aggravates the problems.

Marketing bottlenecks

The handloom industry suffers from marketing-related issues because of poor financial and managerial resources. Handloom products are not easily available and beyond the reach of common people due to its high price. There is also problem of genuineness along with limited designs of the handloom products available in the markets. International market is not exploited properly due to the limited access and funds.

Lack of modernization

The handloom industry has been using age-old technologies and looms, resulting in low productivity. The continuous and repetitive movements during the running of loom along with unhealthy working atmosphere and use of hazardous chemicals during the production process adversely affect the health of weavers and allied workers in various ways such as body pain, pulmonary problems, chronic bronchitis, eye strain, back pain, headache, etc.

Migration to other fields

According to Fourth All India Handloom Census in 2019–2020, the majority (66.3%) of the weaver households earn less than Rs. 5000.00 per month. This relatively low income and unstable work may likely be the key factor behind the fall in average number of weavers per household to 1.05 as compared with 1.28 in the third census. The young generation of weavers has been migrating to other occupations due to this lower income level.

Poor infrastructure

At an aggregate level, only 18.4% of handloom weavers live in pucca structures whereas this figure is 61.9% for Indian family in general. The handloom weaving is carried out in the weaver households spread over a vast geographical area. A weaver household should have at least one weaver, but may or may not have any allied workers. The weaver household lacks the necessary infrastructure in terms of separate sheds, water and power supply, technology support, effluent treatment plants and waste management arrangements. This poor infrastructure adversely affects the productivity, quality and cost.

Lack of reliable data

Lack of reliable data with respect to a number of weavers and allied workers, their socio-economic and livelihood conditions, family details and their productivity is a major short coming that affects the planning and policy formation of handloom sector. This insufficient and lack of reliable data hinders the growth of handloom sectors due to non-possibility of inter-sectoral comparison.

The handloom sector generates sustainable livelihood for a large population, hence reliable data are needed to assess the requirements of the weavers and also to understand the impact of policies on the sector. In this context, Ministry of Textiles has come out with a report on Fourth All India Handloom Census in 2019–2020 involving 31 States and Union Territories practicing the art of weaving. This census serves a dual purpose by creating an updated database of the weavers, allied workers and non-households; and distribution of *Pehchan* cards along with yarn pass books to the weavers. A brief extract from the report on Fourth All India Handloom Census is mentioned hereunder [4].

The handloom weaving is carried out with workforces contributed by the entire family. Across India, 28.2 lakh handlooms are reported in the Fourth All India Handloom Census [4], out of which 25.2 lakhs are in rural areas and 2.9 lakhs located in urban areas. 95.6% of handlooms are located in handloom weaver households, which clearly signify that weaving on handloom is primarily a household-based activity. In rural areas, 41.7% of the looms are pit looms as compared with 53.1% in urban areas, whereas 31.8% are frame loom in rural areas and 30.2% in urban áreas [4]. Loin looms are not very popular in urban area and in rural area 15.3% are loin loom. The lack of capital/funds and inadequate market demand for handloom products may primarily be responsible for almost 17.6% idle looms in the weaver households and 5.3% in master weaver household [4].

The total number of handloom worker households in India is 31.44 lakhs and 88.7% weaver households are located in rural areas, while 11.3% are in urban areas. The average number of weavers per household is 1.05. The value of average number of weavers per household in the rural areas is 1.04 as compared with 1.10 in urban areas, indicating that there is scope of earning opportunities from weaving in urban areas. Out of the total numbers of weaving households, 10,456 are master weaver households. Assam, West Bengal, Manipur and Tamil Nadu account for 18 lakhs of weaver households in the country. 74.8% of weaver's household belong to Hindu, followed by Muslim (16.4%). At an aggregate level, 31.3% of the weaver's household belong to general caste, 34.6% other backward class and 34% SC and ST. The average number of person-days of engagement in weaving activity is 208 days. These engagement activities are more for weavers residing in urban areas (262 days) as against those in rural areas (201 days). In the rural area, women weavers engaged in the profession of handloom weaving as a part-time endeavour, whereas 75.6% of the male weaver engaged in weaving activities is of full-time nature. There are significant differences in the nature of engagement between male and female weavers. The female weavers prefer to work independently, whereas almost one-fourth of male weavers prefer to work under master weaver.

Nearly 25% of weavers have not received any formal education, while a further 14% have not even completed primary level. Goa, Maharashtra, Manipur, Telangana and Himachal Pradesh have the most educated weavers, i.e., completed their high school or more. Only 23.4% of weavers have a bank account. The banking penetration among the weavers is higher in urban households (41.8%) than those living in rural areas (20.8%). Insurance penetration is also very low (3.8%) among weaver households.

The allied activities can be categorized as pre-loom and post-loom and are one of the key inputs to complete a finished weaving product with marketable quality. The pre-loom activities consist of winding, warping, sizing, dyeing, loom setting and manual card punching, whereas calendaring has been considered as a post loom activity. An allied worker is someone who undertakes only pre-loom and/or post loom activities. Allied workers are found in both weaver households as well as in allied worker household. According to the Fourth All India Handloom Census (2019–2020), there are approximately 8.5 lakhs of allied workers and 82.2% of them live in rural areas.

The non-household establishments in the handloom sector also contribute to handloom production in many ways. Non-household units are those that have a shed/area and have looms/weaving/allied work being performed within that premises. The majority of non-household establishments (70.3%) are co-operative societies, followed by units run by others (13.6%). This trend is very similar across rural and urban areas.

In the handloom sectors, the largest yarn type used by handloom weavers in India is cotton of different counts depending on the type of products. The yarn is procured in hank form by the weavers and/or master weavers from the local market, NHDC, local yarn dealers, and Mahajan either in the grey form or dyed form. The yarn in the grey form requires further treatments viz. scouring and bleaching before dyeing. Generally, bleaching is done with bleaching powder. This chlorine-containing bleaching agent is considered as highly toxic and is the main source of AOX, i.e., absorbable organo-halogen compounds [5]. Dyeing is one of the most important value-addition processes in the handloom sector. The dyeing process is either carried out in the yarn form before selling the yarn or on fabric. 60% of the yarn is sold in the grey form and 40% in dyed form [4]. The main classes of dyes that are commonly used in handloom sector for processing cotton materials are direct, vat, naphthol and reactive. However, in the recent years, there has been a shift in interest towards natural colourants obtained from vegetable resources. The entire processes are performed by the dyer in their home or separate premises using crude methods. The process parameters viz. temperature, time, pH, quantities of dyes and chemicals are not controlled properly leading to inferior product quality. The loss in strength of the yarn after processing adversely affects the productivity and quality of the end products. Silk is the second preferred yarn in the handloom sector and among the silk, the most popular yarn is muga and mulberry, followed by eri. There are also market demands for the fabrics/products made out of wool, linen, polyester blends and acrylic. The majority of the fabrics produced in the handloom sectors across the country depends on the choice for which the fabrics are to be considered. The fabrics produced in handloom sectors are saree (22.9%), shawls/mekhla chadder/stole/scarf (26.7%), dhoti/lungi/angavastra (19.5%), towel, napkin. gamcha (16.5%), durries, rugs, mats (3.5%), dress materials (3.2%), bedsheet, furnishing, blankets (3.4%). The cloth woven in every part of the country is the distinct specialization of those geographical areas [4].

2 Need for Value-Addition

The handloom industry has innumerable items to offer and the element of art and craft makes it a potential sector for the upper segments of domestic and global market. The vulnerable financial conditions of the weavers restrict them to spent money on design development in accordance with the changing taste of the consumers. Hence, value-addition and product diversification are very much essential for the survival of this rich cultural heritage of the country. Silpa-Sadana, a department under Palli Samgathana Vibhaga of Visva-Bharati (A Central University) had taken a leading role in revitalizing the decadent rural industries and craft sector in India, and also a premier institute working in the field of handloom textiles. There is an opportunity to add value to the handloom textiles utilizing different types of yarns viz. eri, organic cotton, linen etc., design innovation, application of natural dyes for dyeing and printing, surface ornamentation by batik work etc. This may improve the livelihood of handloom weavers and also enrich the aesthetic appeal of the end products. This sector has a great deal of potential for further value addition in the Ready-Made Garment (RMG) sector. It is also very important to create awareness to the younger generation about the importance of handloom industry and its contribution to the socio-economic development of the country. The younger generation should be attracted to the handloom products through design innovation as per present market trend, style and comfort.

3 Value-Added Handloom Textiles

3.1 Dyeing of Eri with Natural Colour

Eri silk is the only completely domesticated non-mulberry variety and eri culture is a household activity by the tribal of northeastern part of our country. It is a multivoltine silk and not composed of continuous filaments. Eri cocoons (Fig. 1) are degummed prior to spinning [6]. The degumming process removes the gum partially, which facilitates yarn formation. The demand for both handspun and mill spun eri yarn is gradually increasing in the domestic and export market.

3.1.1 Extraction of Natural Dye

In the market, natural dyes are available in crude form, powder form and paste form. The process of extraction can be eliminated in case of paste and water extracted powder form. In case of crude form, the extraction process with solvent is required and commercially water is used as solvent. Before extraction, the vegetable matters are dried in the absence of direct sunlight, followed by grinding to break down the materials into powder form (Fig. 2).

Fig. 1 Eri cocoon

Fig. 2 Crushing of vegetable matters **a** crude form, **b** mechanical crusher, and **c** powder form

Fig. 3 Extraction process **a** thermostat control water bath, **b** extracted solution, **c** filtration and **d** aqueous solution of dye

The extraction of vegetable matters is performed by adding specified quantities of those crushed matters to a fixed volume of water. Generally, the extraction process is either performed in room temperature or at boil. It is always advisable to maintain a constant temperature throughout the extraction process. The extraction at boil is carried out for 45–60 min in a thermostat control water bath (Fig. 3) and finally filtered through nylon bolting cloth to separate the colouring components and roughages. The evaporated water is replaced by fresh water in order to maintain the pre-established concentration of the dyed solution and used for dyeing and printing purposes.

3.1.2 Dyeing of Eri Yarn with Natural Dye

Application of aqueous extract of natural dyes viz. *Rubia cordifolia L*, *Laccifer lacca Kerr* and *Allium cepa L* is done following post-mordanting methods. Inorganic salts viz. aluminium sulphate and ferrous sulphate are used as mordanting agent. The *Laccifer lacca Kerr* dye used for dyeing of eri yarn is purchased from the market in paste form. In case of post-mordanting method, the dyeing is carried out at 80 °C for 20 min in a thermostat control beaker dyeing machine with continuous stirring. The *p*H of the dye bath is kept at ~4.5 with the addition of acetic acid (5 ml/l). After dyeing, the yarn is treated with either aluminium sulphate (10 g/l) or ferrous sulphate as per the requirement of shades. The mordanting process is carried out at 70 °C for another 20 min. After mordanting, the yarn is washed in cold water, followed by

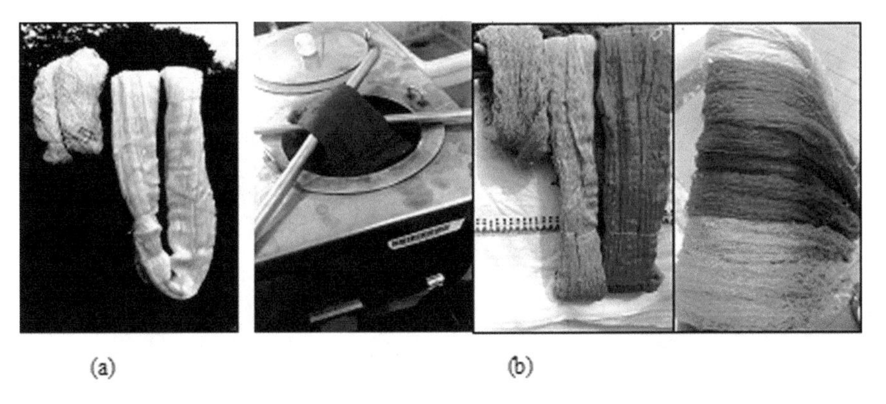

Fig. 4 **a** Undyed eri yarn and **b** eri yarn dyed with natural colourants. *Courtesy* Roslin Akhtar Hussain, Major Project, Bachelor in Design (Specialization in Textiles) under the supervision of Dr. Sankar Roy Maulik 2020

soaping with non-ionic detergent (2 g/l) at 50 °C for 5 min in order to remove the unfixed dye from the yarn surface. Finally, scrooping of the dyed yarn (Fig. 4) is performed with the help of acetic acid (5 ml/l) at room temperature for 5 min.

3.1.3 Dyeing of Eri Yarn with Indigo

The plant *Indigofera tinctoria* belongs to the family Fabaceae is a shrub of around 1.2–1.8 m in height, and it is one of the oldest dyes used for dyeing and printing of textiles. The process of dyeing eri yarn with indigo requires precautions. The Indigo vat is prepared using Indigo, sodium hydrosulphite and sodium hydroxide. The vatting and solubilising process is performed at room temperature for 10 min. Eri yarns are carefully dipped in this vatted solution for 20 min. The dyed Eri yarns are dipped in the solution containing hydrogen peroxide and acetic acid for 10 min. Soaping of the dyed yarn is done employing non-ionic detergent for 5 min at 50 °C. Finally, scrooping of the dyed yarn (Fig. 5) is performed with the help of acetic acid.

3.2 Printing and Painting with Natural Colour

Textile printing is used for introducing colour and design to the fabrics. It is the process of bringing together a design idea, colourants and textile substrate. Printing on textile is a very skilful process and the printed fabrics have much more aesthetic appeal than the dyed varieties [7]. India has a rich heritage and tradition of decorating textiles through printing or painting, and colourants extracted from natural resources were used by the painters to decorate their paintings [8, 9].

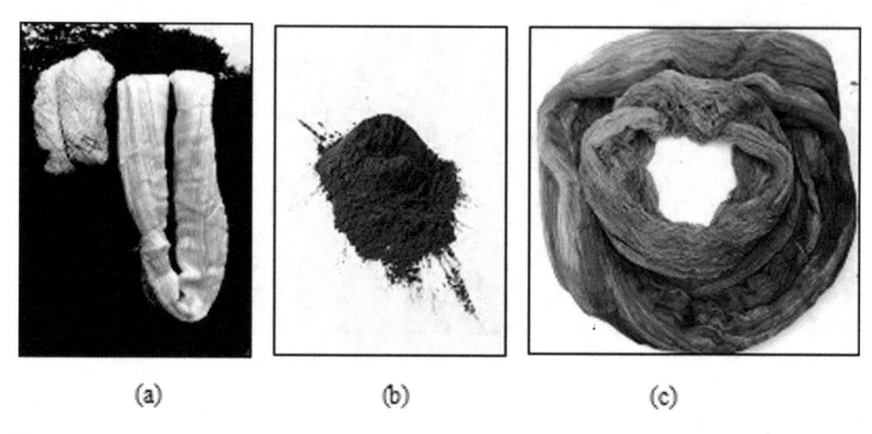

Fig. 5 **a** Undyed eri yarn, **b** Indigo and **c** eri yarn dyed with natural indigo. *Courtesy* Roslin Akhtar Hussain, Major Project, Bachelor in Design (Specialization in Textiles) under the supervision of Dr. Sankar Roy Maulik 2020

3.2.1 Fabric Preparation Before Printing and Painting

Degumming of Silk

Degumming of silk fabric is performed at 90 °C for1.5 h in an aqueous solution containing olive oil-based soap (6 g/l) and sodium carbonate (2 g/l) at a fabric-to-liquor ratio of 1:20 in a thermostatically controlled water bath (Fig. 3a). Degummed fabric is washed at 70 °C for 10 min, cold washed and finally dried in air.

Desizing of Cotton Fabric

Desizing of handloom cotton fabric is performed using 0.25(N) hydrochloric acid solution at 50 °C for 2 h, at a fabric-to-liquor ratio of 1:20. The desized fabric is washed thoroughly using hot water, followed by a cold wash prior to combined scouring and bleaching process.

Combined Scouring and Bleaching of Cotton

Combined scouring and bleaching of the desized cotton fabric are performed in an aqueous solution containing sodium hydroxide (3%), sodium carbonate (3%), anionic detergent (1%), Turkey Red Oil (1%) and sodium silicate (2%) at boil for 2 h keeping a material-to-liquor ratio at 1:20. At the time of boiling, hydrogen peroxide solution (2%) is added in two instalments and the process continues for another 1 h. The scoured and bleached cotton fabric is washed with hot water, followed by cold wash and neutralized with dilute acetic acid, and finally dried in air [10].

3.2.2 Printing and Painting with Natural Colour

Dyeing of silk and cotton fabrics with the aqueous extract of natural colourants is performed in the absence and/or presence of different inorganic salts or mordants either through exhaust method or padding process. The padding is performed in a two-bowl padding mangle at 100% wet pick up and after padding, the fabrics are dried at room temperature [11] followed by printing. The printing/painting paste is prepared by adding inorganic salts or mordants of a specified dose level with the aqueous extract of natural colourants. This mixture is kept for 15–20 min in order to form lake or complex, followed by addition of natural gum with the help of high-speed stirrer for preparing the printing/painting paste. The printing/painting on silk and cotton fabrics is performed with the help of wooden blocks of various designs, brush etc. (Figs. 6, 7 and 8). After printing/painting, the fabrics are dried at room temperature, followed by steaming at 102 °C for 30–45 min in a cottage steamer (Figure 9). After steaming, the printed/painted fabrics are immediately washed in a solution containing 2 g/l non-ionic detergent at 60 °C for 10 min, followed by washing with cold water and finally dried in air [12].

Discharge Style of Printing

Cotton fabric is padded with an aqueous solution of *Terminalia chebula* and ferrous (iron) sulphate by the help of hand-driven padding mangle. The dyed fabric is then painted with a paste consists of oxalic acid, ammonium hydroxide, sodium alginate and water, followed by steaming at 102 °C for 15 min. The fabric is then washed

Fig. 6 Hand-painted silk stole. *Courtesy* Rafiya Tunnesa Begam, Major Project, Master of Design—Textiles under the supervision of Dr. Sankar Roy Maulik 2019 (Model: Sayani Banerjee)

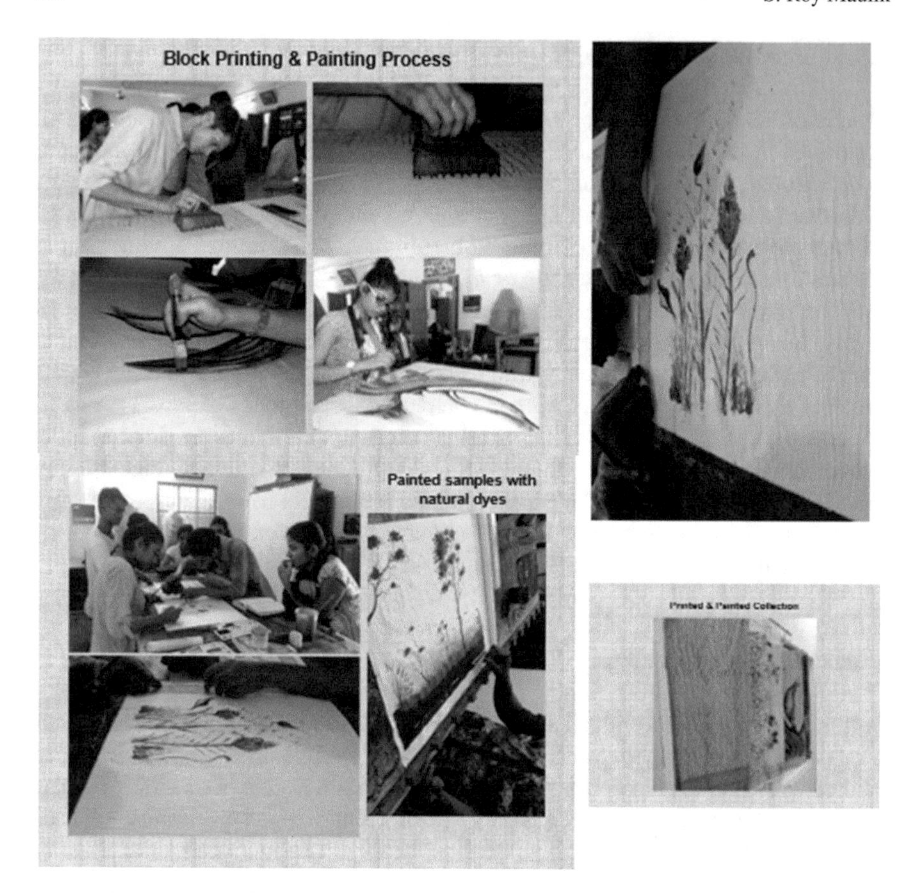

Fig. 7 Block printing and hand painting on cotton and silk fabrics. *Courtesy* Students of Bachelor in Design (Specialization in Textiles and Clothing) under the supervision of Dr. Sankar Roy Maulik

thoroughly with 2 g/l non-ionic detergent, followed by cold wash and finally dried in air (Fig. 10).

3.3 Batik Work with Natural Colour

Batik is an old craft and one of the most popular surface ornamentation techniques on textile to achieve value and aesthetic appeal. The Indonesian Batik is best known for its intricacy and aesthetic look. Rabindranath Tagore went to Java in 1927 and noticed this exquisite art. The revival of Batik in India began at Santiniketan during the twentieth century [13]. In early days, Batik work with natural colours was very common, but presently dyes obtained from petrochemical sources are used in Batik work due to its easier application process and wider colour range. The present market

Fig. 8 Natural dyed and printed handloom textiles. *Courtesy* Manisha Priya, Major Project, Bachelor in Design (Specialization in Textiles and Clothing) under the supervision of Dr. Sankar Roy Maulik 2019

demand has forced the artisans to use synthetic dyes in place of natural colour for Batik work. It is preferable to use dyestuffs, which can be applied in cold or room temperature. Among the various class of synthetic dyes, naphthol colour, solubilized vat dye and remazol class of reactive dyes are mostly used in Batik work. Batik work offers immense possibilities for artistic freedom as patterns are applied by actual drawing. The batik has gained popularity not only in apparel but also in furnishing, heavy canvas, wall hangings, table cloths and household accessories. The increasing global awareness on environmental issues and aesthetic values has revived interest of using natural dyes in Batik work, and it has opened up opportunities in domestic and international market.

3.3.1 Fabric

The Batik work is mainly done on natural material viz. cotton and silk fabric due to its good absorbing power. Removal of natural and added impurities from those fabrics is very important before Batik work. In early days, traditional methods of removing the impurities from the natural fabrics were followed in order to make the cloth more absorbent. But presently, conventional process of removing the impurities is more

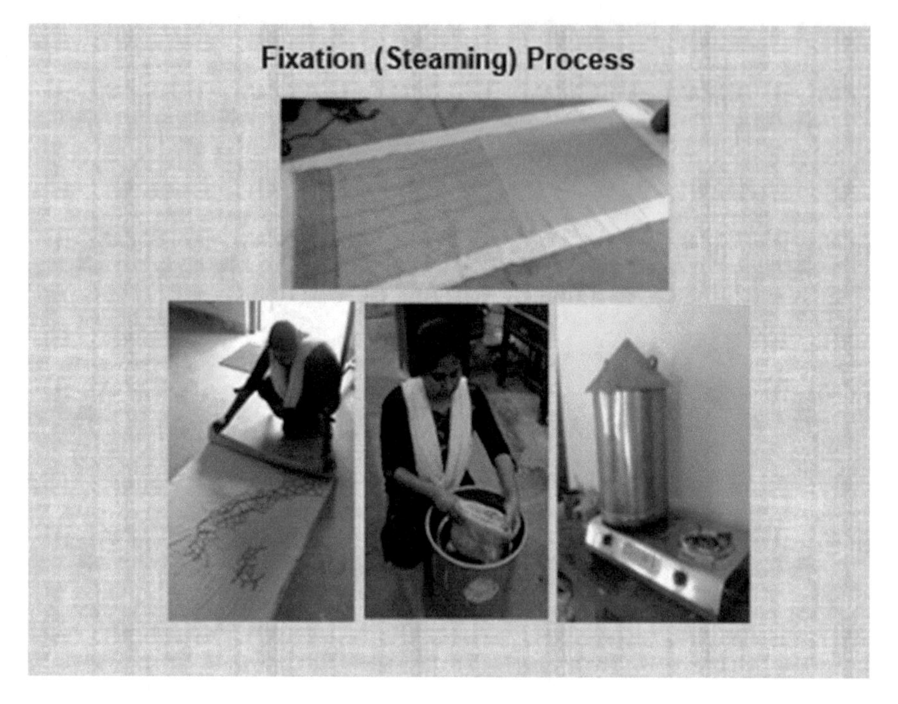

Fig. 9 Steaming process for fixation of the dye. *Courtesy* Rafiya Tunnesa Begam, Bachelor in Design (Specialization in Textiles and Clothing) under the supervision of Dr. Sankar Roy Maulik 2015

beneficial in terms of cost, time and quality. The process of removing impurities from cotton and silk fabrics has already been reported in Sect. 3.2.1.

3.3.2 Resisting Agents

Beeswax and paraffin wax are used to impede the colour into the selected areas during dyeing. Beeswax is soft and acts as a strong shield against the dye. On the other hand, paraffin wax is hard and brittle and mainly used for creating cracks. Generally, paraffin wax and bee wax are used in Batik work in 1:1 ratio, but 100% paraffin wax can also be used [14]. Sometime, a natural resinous substance is also mixed along with wax in a small quantity.

3.3.3 Batik Work

It is a process of dyeing the fabric by resisting the specified area with wax. Batik work with natural dyes can be done in the following steps [14].

Fig. 10 Discharge style of painting. *Courtesy* Rafiya Tunnesa Begam, Bachelor in Design (Specialization in Textiles and Clothing) under the supervision of Dr. Sankar Roy Maulik 2015 (Model: Sanjukta Acherjee)

Design

The drawing is traced out on the fabric as per design. The design is traced on tracing paper and needle is used to create holes along the outline of the design. The fabric is then placed on a hard and plain surface and the tracing paper is attached to the fabric using pin, and the design is transferred to the fabric with the help of kerosene and blueing agent. The tracing paper is then removed carefully to get the final design on the fabric.

Waxing

It is a process of applying wax over the drawing areas according to the design (Fig. 11). The waxing process is done in steps from the lightest to the darkest one.

Fig. 11 Batik work. *Courtesy* Kazi Md. Nasiruddin, Master of Design—Textiles, and Manisha Priya, Bachelor in Design (Specialization in Textiles and Clothing) at IIHT, Guwahati under the supervision of Dr. Sankar Roy Maulik 2019

Dyeing

Batik is mainly done by the artisans and small enterprises. This industry causes severe water pollution due to the use of naphthol colour and other restricted chemicals. Hence, there is a need for an alternative eco-friendly route in order to sustain this rich cultural heritage. Batik work on handloom fabric with natural dye (Fig. 11) may be one of such alternatives to preserve the heritage of our country. The dyeing of wax-coated fabric with aqueous extract of natural colourants can be carried out either following post- or simultaneous mordanting methods. In simultaneous mordanting, appropriate quantity of mordant is mixed with the aqueous extract of natural colourants and kept for 15–30 min at room temperature. The wax-coated fabric is then dipped in this solution for 15–20 min at room temperature, followed by drying. In post-mordanting method, the wax-coated fabric is impregnated into the aqueous extract of natural dye, followed by treatment with mordant. The process of waxing followed by dyeing is repeated on the same fabric with different colours to create elaborate and vibrant designs.

Dyeing with Indigo

Indigo is very popular dye for Batik work (Fig. 12). The wax-coated fabric is dipped into a solution comprising of *Indigofera tinctoria,* sodium hydrosulphite and sodium

Fig. 12 Batik work with *Indigofera tinctoria*. *Source* Sankar Roy Maulik and Lina Chakraborty, Batik works with *Indigofera tinctoria*, *Indian Journal of Natural Fibres*, 2(2), 2016, pp 35–39

hydroxide, and dyeing process is carried out in accordance with the method described in Sect. 3.1.3.

Wax Removal

Wax should be removed properly after completion of the Batik work. Natural dyes are very much sensitive to alkali, hence conventional wax removing process is not possible for this colour. In case of natural dyes, non-ionic detergent and emulsifying agent are used for removing wax. Removal of wax produces a contrast between the dyed and resisting portions.

3.4 Eco-Friendly Woven Apparels

Green consumerization is a growing phenomenon nowadays and the purchasing decision of a customer is based on style, comfort, environmental impacts and product

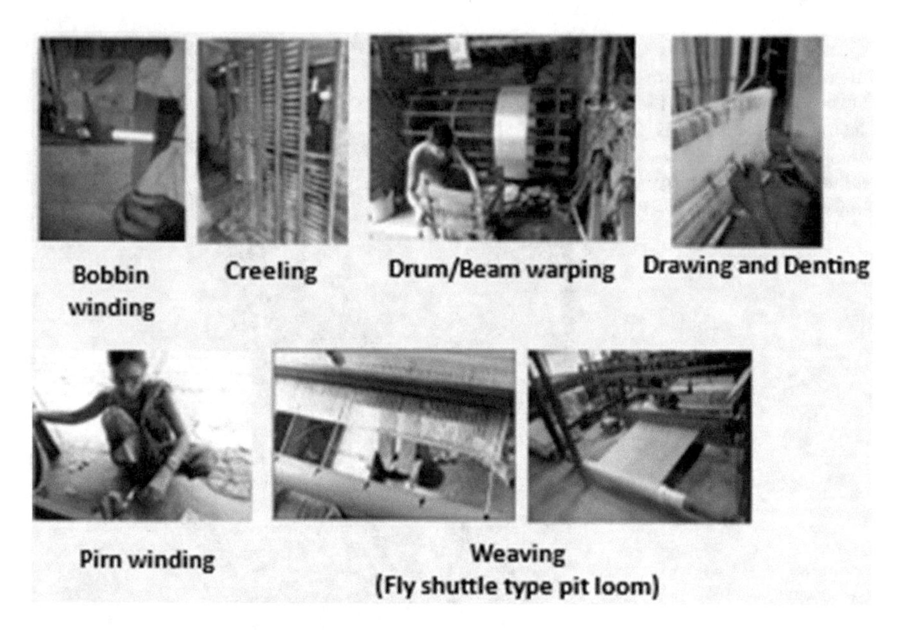

Fig. 13 Different stages of handloom fabric formation. *Source* Lina Chakraborty and Sankar Roy Maulik, International Conference on "*Redefining Textiles—Cutting Edge Technology of the Future (RTCT-2016)*" organized by Dr. B R Ambedkar National Institute of Technology, Jalandhar, during 8–10 April 2016

safety requirements. The consumers are more aware about the safety of chemicals and their impact on environment. In this context, there is a worldwide interest in using natural dyes for colouration on textiles. The eco-friendly apparels made from handloom fabrics may create a new horizon in the niche market for upper and middle segment of the society. This value addition may enrich the aesthetic appeal of the end product and attract the new generation while purchasing handloom textiles.

The handloom fabrics are produced in fly shuttle type pit loom through various preparatory steps viz. bobbin winding, beam warping, drawing and denting, looming, pirn winding etc. prior to weaving [15]. In all the cases, scoured/natural dyed cotton is used as warp yarn, whereas eri, cotton and spun silk dyed with natural dyes are used as weft. The fabrics are produced according to the design and the various stages of fabric production process are described in Fig. 13.

4 Conclusion

The handloom sector represents the economic lifeline of the most vulnerable sections of our society. Certain traditional handlooms appear to be slowly dying due to a lack of awareness and inadequate appreciation of the intricacies and skills involved. However, the growing competitiveness both in the national and international markets,

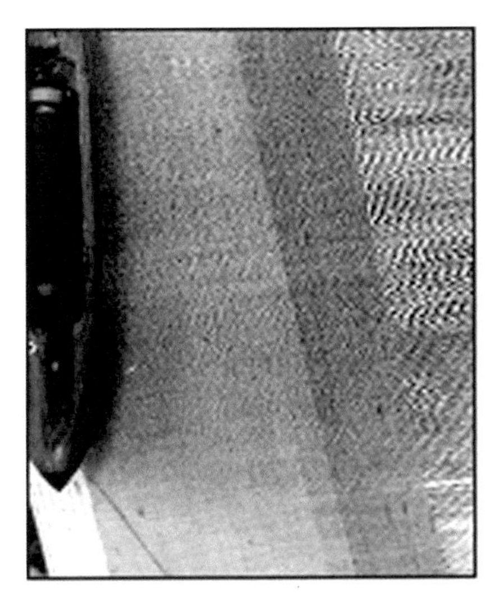

Fig. 14 Plain woven union fabric made with cotton and eri silk. *Courtesy* Roslin Akhtar Hussain, Major Project, Bachelor in Design (Specialization in Textiles) under the supervision of Dr. Sankar Roy Maulik, 2020

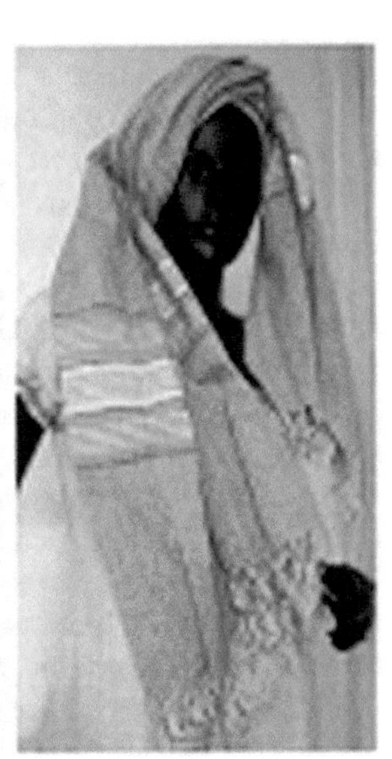

Fig. 15 Natural and eco-friendly apparel made from handloom fabrics. *Source* Lina Chakraborty and Sankar Roy Maulik, International Conference *"Emerging Trends in Traditional and Technical Textiles"* organized *by* Department of Textile Technology, Dr. B R Ambedkar National Institute of Technology, Jalandhar, during 11–12 April, 2014

Fig. 16 Value-added apparel made of handloom textiles. *Source* Lina Chakraborty and Sankar Roy Maulik, International Conference on "*Redefining Textiles—Cutting Edge Technology of the Future (RTCT-2016)*" organized by Dr. B R Ambedkar National Institute of Technology, Jalandhar, during 8–10 April 2016

it is the need of the hour to adopt a focused and holistic approach in the handloom sector to meet the challenges of a globalized environment. The concept of sustainable development due to the growing consciousness of environmental issues along with other socio-economic problems has forced the industries to develop a sustainable production system. The ever-changing consumer preferences also force the industry to redefine their business strategies. The value-addition of handloom textiles has been identified as one of the major thrust areas for the overall development of the sector. It is, therefore, very essential to explore ideas through design innovation and using eco-friendly dyes for the survival of this rich cultural heritage. Natural dyed fabric is gaining momentum due to the fast-changing fashion trend. Designers, manufacturers and retailers are busy to develop 'green' product range for the mass market and organic non-toxic products in every field have created a new horizon. Since handloom itself is an eco-friendly process of manufacturing fabrics and hence the use of natural dyes in the presence or absence of different innocuous inorganic salts really completes sustainable textiles. Making handloom fabric popular will give a fresh lease of life to this dying art, and people involved in this industry will be able to use their skills to maintain their families while keeping alive the art form that would otherwise soon become a thing of the past. Cluster-based approach and value-addition of handloom products by utilizing different yarns apart from cotton and using eco-friendly dyeing/printing/surface ornamentation etc. may be few approaches to revitalize and also to tap the full potential of this traditional heritage.

References

1. Annual Report, Ministry of Textiles, Government of India (2018–2019)
2. Implementation of Cluster Development Initiatives in Shantipur Handloom Textiles Cluster, under IHCDS, Development Commissioner (Handloom), Government of India, Textile Committee, Ministry of Textiles (2014)

3. Behera S, Khandual A (2017) Annual Technical Volume of Textile Engineering Division Board. Instit Eng (India) 2:51–62
4. Fourth All India Handloom Census, Office of the Development Commissioner for Handlooms, Ministry of Textiles, Government of India, (2019–2020)
5. Roy Maulik S (20017) "Eco-friendly processing of textiles", Annual technical volume of textile engineering division board. Instit Engs (India) 2:36–44. ISBN 978–81–932567–7–0
6. Roy Maulik S, Mondal S, Munshi R (2018) Eri cocoon degumming and its influence on yarn properties. Asian Dyer 15(4):61–63
7. Roy Maulik S, Roy F (2011) Printing of cotton fabric with pigment colour. Asian Dyer 8(4):52–58
8. Govindwar S, Grover E, Fatima N, Paul S (2011) Printing of khadi fabric using various natural thickeners and synthetic dyes. Asian Dyer 8(3):49–52
9. Roy Maulik S, Agarwal K (2014) Painting on handloom cotton fabric with colourants extracted from natural sources. Indian J. Trad Knowl 13(3):589–595
10. Roy Maulik S, Mandal S (2010) Printing of handloom cotton fabric with natural colour. Asian Dyer 7(2):49–54
11. Roy Maulik S (2019) Application of natural dyes on protein fibres following pad-steam methods. J Instit Eng (India): Ser E 100(1):1–9. https://doi.org/10.1007/s40034-019-00141-5.
12. Roy Maulik S (2015) Printing of silk fabric following simultaneous mordanting technique. J Text Assoc 75(5):345–50
13. Kirsur MV (2007) Beautiful batiks of Bhairongarh. Indian Silk 3:26
14. Roy Maulik S, Bhowmik L, Agarwal K (2014) Batik on handloom cotton fabric with natural dye. Indian J Trad Knowl 13(4):788–794
15. Roy Maulik S, Bhowmik L, Agarwal K (2016) Eco-friendly woven and printed apparel made from handloom cotton fabric. Asian Dyer 13(6):51–54

Printed in the United States
by Baker & Taylor Publisher Services